AF541092

SUPERFOODS
Modern Trends

SUPERFOODS
Modern Trends

– Author –

Aatreya Challa
Ilanthirayan Chittem

ISBN: 9788119283033 (Hardback)

Published by : **WORDPEN ACADEMICS**
4736/23, Ansari Road, Darya Ganj,
New Delhi – 110 002
Phone: +91-011-35004336
E-mail: info@wordpenacademics.com

Printed at **:** **Replika Press Pvt. Ltd.**

Exclusive Distributor: ***Bio-Green Books, New Delhi***

Preface

Superfoods are nutrient-dense foods that are believed to provide a range of health benefits. These foods are typically rich in vitamins, minerals, antioxidants, and other beneficial compounds that can help to reduce the risk of chronic diseases and promote overall health and well-being. The concept of superfoods has gained popularity in recent years, with many people seeking to incorporate these foods into their diets.The term "superfood" is not a scientific term and is not recognized by the FDA. Rather, it is a marketing term that is used to describe certain foods that are believed to have exceptional nutritional value. Some of the most commonly recognized superfoods include berries, leafy greens, nuts, seeds, fatty fish, and whole grains. These foods are typically rich in vitamins, minerals, and other beneficial compounds such as antioxidants, polyphenols, and fiber. One of the primary benefits of superfoods is their ability to reduce the risk of chronic diseases. Many chronic diseases such as heart disease, cancer, and diabetes are associated with oxidative stress and inflammation. Superfoods are typically rich in antioxidants and other beneficial compounds that can help to reduce oxidative stress and inflammation in the body. For example, berries are high in anthocyanins, which are powerful antioxidants that have been shown to reduce the risk of heart disease and cancer. Leafy greens are high in vitamins and minerals, as well as phytochemicals that have been shown to reduce the risk of chronic diseases.

In addition to reducing the risk of chronic diseases, superfoods are also believed to promote overall health and well-being. These foods are typically nutrient-dense, meaning that they provide a wide range of vitamins, minerals, and other beneficial compounds that are essential for good health. Eating a diet rich in superfoods can help to boost energy levels, improve mental clarity, and promote a healthy weight. It is important to note that superfoods are not a magic bullet for good health. While incorporating these foods into your diet can provide a range of health benefits, they should be part of a balanced diet that includes a variety of foods from all the major food groups. In addition, it is important to choose superfoods that are in season and locally grown whenever possible, as this can help to ensure that they are at their peak of nutritional value. Superfoods are a group of nutrient-dense foods that are believed to provide a range of health benefits. These foods are typically rich in vitamins, minerals, antioxidants, and other beneficial compounds that can help to reduce the risk of chronic diseases and promote overall health and well-being. While incorporating superfoods into your diet can be beneficial, it is important to remember that they should be part of a balanced diet that includes a variety of foods from all the major food groups.

It discusses the design of an ideal diet using common foods and provides an overview of the concept of super foods and their role in a healthy diet. Focuses on the bioactive compounds contained in the Mediterranean diet and their effects on neurodegenerative diseases and provides an in-depth analysis of the health benefits of the Mediterranean diet, which is considered to be one of the healthiest diets in the world. It explores the development and characterization of fish-based super foods and provides information on the nutritional content and health benefits of fish-based super foods. It provides an overview of the bioactive compounds contained in beer brewing byproducts and their potential health benefits, including anti-inflammatory and anti-cancer properties. It describes the methods used to measure bioaccessibility and their importance in determining the bioavailability of nutrients in super foods. The book is likely to appeal to researchers, academics, and health professionals interested in nutrition, diet, and public health.

We are grateful to all those persons as well as various books, manuals, periodicals, magazines, journals etc. that helped in the preparation of this book. In spite of the best efforts, it is possible that some errors may have occurred into the compilation and editing of the book. Further queries, constructive suggestions and criticisms for the improvement of the book are always welcomed and shall be thankfully acknowledged.

Aatreya Challa
Ilanthirayan Chittem

Contents

Preface .... *v*

1. **Introductory Chapter: Design of an Ideal Diet Using Common Foods** **1**
    - Introduction
    - Superfoods generally used in Japanese cuisines
    - Ideal diet using common foods containing active ingredients
    - Conclusion
    - References
2. **Bioactive Compounds Contained in Mediterranean Diet and Their Effects on Neurodegenerative Diseases** **13**
    - Introduction
    - Bioactive substances in the Mediterranean diet
    - Conclusion
    - References
3. **Development and Characterization of Fish-Based Superfoods** **33**
    - Introduction
    - Recovery of proteins from fish processing by-products
    - Nutritional properties of recovered proteins by isoelectric solubilization/precipitation
    - Assessment of impact of ω-3 PUFA incorporation on fish-based protein for superfoods
    - Role of salt substitute and ω-3 PUFAs rich oil in fish-based superfoods
    - Physicochemical properties of fish-based superfoods fortified with dietary fiber
    - Conclusions
    - References

4. **Health Potential for Beer Brewing Byproducts 49**

- Introduction
- Brewing beer and brewing byproducts
- Composition and current applications of beer brewing byproducts
- Functional components in beer brewing byproducts
- Health benefits and potential of byproducts from beer brewing
- Conclusion
- References

5. **Assessment of Bio-accessibility: A Vital Aspect for Determining the Efficacy of Superfoods 67**

- Introduction
- Stability of the antioxidant and starch hydrolase inhibitory activities of five Sri Lankan fruits subjected to pancreatic and duodenal digestion
- Evaluation of the antioxidant and starch hydrolase inhibitory activity of 10 commonly used household spices subjected to in vitro digestion
- Stability of the antioxidant and starch hydrolase inhibitory activities of Kombucha teas prepared from three microbial cultures
- Conclusions
- References

Index 81

Chapter 1

Introductory Chapter: Design of an Ideal Diet Using Common Foods

1. Introduction

In recent times, lifestyle-related diseases, such as diabetes and heart disease, are increasing due to changes in lifestyle and dietary habits. Rapid increase in type 2 diabetes is becoming a serious problem, especially in South Asian countries [1]. There is also an increase in the number of patients with cancer and dementia worldwide. In addition, the elderly population in the world is increasing, while dementia is becoming a serious social problem in Japan. The risk associated with these diseases can be decreased by improving eating habits, and the best way is to control total calorie intake and eat foods containing active ingredients to prevent these diseases.

Many active ingredients in food to prevent these diseases have been discovered a long time ago, and effective consumption methods for dairy food have been developed. For instance, active ingredients extracted from food or artificially synthesized are found in the form of tablets for use as a dietary supplement. Many people living in Europe and the United States prefer dietary supplements, while the Japanese people prefer food-type to tablet-type supplements. Consequently, "functional foods" have been developed that comprise artificial foods in which active ingredients are added [2]. Currently, various functional foods are developed and sold in Japan.

Recently, superfoods, instead of dietary supplements or functional foods, are considered an effective mode of intake of active ingredients. The word "superfood" is a marketing term and is not correctly defined. The idea of superfood was first proposed by Platt [3] and Wolfe [4]. According to the concept of Platt, superfood can be defined as a low-calorie food containing active ingredients that decrease the risk of diseases, such as cancer and dementia. Wolfe defined the word "superfood" as a food that contains extremely high concentrations of active ingredients or important nutrients. Foods shown in **Table 1** have been selected as typical superfoods based on these concepts, and their characteristics have been previously examined [5, 6].

Superfoods	Active ingredients or important nutrients at extremely high concentrations
Açaí	Anthocyanin
Black Garlic	Cycloalliin and S-allyl-L-cysteine
Broccoli (Super Sprout)	Sulforaphane
Cacao/Cocoa powder	Cocoa polyphenols
Camu camu	Vitamin C
Chia seeds	Glucomannan
Goji berries	β-Carotene
Hemp seeds	α-Linolenic acid
Maca	Minerals (iron and calcium)
Mangosteen	Xanthone
Coconut oil	Lauric acid (Medium Chain Triglyceride)
Spirulina	Various nutrients

Table 1. Typical superfoods and their respective main active ingredient or important nutrients.

On the other hand, several common foods consumed on a daily basis contain sufficient amounts of active ingredients, and therefore, they should also be considered "superfoods." A diet with a well-balanced combination of those foods is the most effective method to consume active ingredients because the ingredients are easily available, used on a daily basis, and can be used in several different recipes. However, there are few reviews on this approach. Thus, in the introductory chapter of this book, I introduced the concept of an ideal diet using common superfoods. I recommend that the interested readers read the other chapters of this book on the development of superfoods because the purpose of this book is to introduce both the current topics on the development of superfoods and diet using superfoods.

2. Superfoods generally used in Japanese cuisines

Before discussing an ideal diet using common foods, I introduced the active ingredients included in common foods that decrease the risk of lifestyle-related diseases, such as cancer and dementia. I selected foods that are consumed on a daily basis in Japan, with active ingredients that have been precisely examined previously.

2.1. Foods that decrease the risk of lifestyle diseases

Type-2 diabetes, high blood pressure, hyperlipidemia, and heart failure are the common lifestyle-related diseases in Japan. These diseases can occur in conjunction with each other, like a domino effect, and sometimes, more than two diseases can often occur simultaneously, and is referred to as the "metabolic syndrome." The main causes of the metabolic syndrome are insulin resistance and high glucose consumption induced by obesity [7, 8]. Therefore, reducing obesity is the best countermeasure to decrease the risk of lifestyle-related diseases.

The choice of oil is most impactful in reducing obesity because fatty acid composition strongly influences the triglyceride characteristics. Docosahexaenoic acid (DHA) and eicosapentaenoic acid (EPA) of ω-3 polyunsaturated fatty acids (PUFAs) are present in high concentrations in fish such as Pacific saury, sardine, mackerel and eel, which are popular fish in Japan. DHA is most effective in enhancing the fluidity of fat. Rapid degradation of fat can be achieved by positive and frequent eating of these fish, consequently reducing obesity [9]. Oleic acid has an effect similar to DHA because it increases the ratio of ω-3/ω-6 by decreasing the number of ω-6 fatty acids [10, 11]. Olive oil contains high concentrations of oleic acid. Since people living in the Mediterranean region frequently use olive oil, the risk of lifestyle-related diseases is lower than that in people living in other European regions [12].

Positive consumption of compounds that can inhibit the absorption of triglycerides or enhance the secretion of cholesterol is also effective in reducing obesity. Triglycerides are degraded to glycerol and fatty acids by lipases at the small intestine. To inhibit absorption, compounds can interact with those fatty acids in the small intestine. Calcium, which is mainly contained in milk, small fish, and soybean, converts fatty acids to fatty acid calcium salt and inhibits absorption. A study on rats showed that absorption of triglycerides could be decreased if high concentrations of calcium are ingested with triglycerides [13]. Intake of polyphenols, such as quercetin, is also effective in inhibiting absorption of triglycerides [14, 15]. Catechins contained in green tea [16] can inhibit both intake and degradation of glucose, thereby decreasing the levels of glucose, total cholesterol, and LDL in the blood. Alginate, a component of seaweeds such as *Wakame* and *Kombu*, promotes release of cholesterol from the body by enhancing of the secretion of bile. Taurine, which is present in octopus and shellfish, is a feedstock of bile and enhances the production rate from cholesterol to bile. Therefore, positive and frequent eating of foods containing these compounds decrease triglyceride and LDL cholesterol levels, consequently preventing arteriosclerosis [17, 18].

Degrading triglycerides or inhibiting absorption is effective in reducing obesity. Fatty acids are degraded via two pathways: (1) to produce energy in the mitochondria and (2) to produce glucose in the case of low blood glucose levels. In the former, fatty acids must be transported, assisted by carnitine, to the inner membrane of the mitochondria before degradation. Several studies have suggested that positive consumption of carnitine enhances the degradation of triglyceride [19]. As carnitine is contained in fish muscle, consumption of fish is the best way of consuming this ingredient without unfavorable intake of animal fat.

Meanwhile, recent studies have estimated that increase or activation of brown adipocytes (or bright adipocytes) is the most effective way to enhance degradation of triglycerides. Uncoupling protein 1 (UCP1) in brown adipocytes is activated by noradrenaline and produces heat by decreasing the potential energy of the inner membrane of mitochondria. The energy decrease results in the degradation of fatty acids via the TCA cycle to reconstitute the energy gradient. Several studies have shown that obesity and type-2 diabetes reduced by the enhancement of UCP1 in humans [20–22]. Capsaicin and shogaol, which are spicy ingredients of red pepper and ginger, respectively, can activate UCP1 via the TRPV1 receptor, producing heat. Therefore, continuous intake of these ingredients reduces obesity [23].

2.2. Foods to prevent aging

Recent studies have elucidated the mechanism of aging. One important factor that accelerates aging is oxidation of lipids, proteins, and telomeres by reactive oxygen species (ROS). Superoxide dismutase and catalase enzymes degrade ROS from mitochondria, preventing the harmful effects of ROS. The activity of these enzymes in the elders is much lower than those in the young people. Therefore, the increase in oxidation in lipids, proteins, and DNAs in the elderly results in aging [24]. Antioxidants are effective in preventing the harmful effects of ROS. Vitamin C, vitamin E, and β-carotene are known antioxidants; many polyphenols and other compounds have been discovered as novel antioxidants with increased activity [25, 26], for example, catechin (component of green tea), anthocyanin (berries), daidzein and genestin (soybean), sesaminol (sesame), and lycopene (tomato). Eating foods containing sufficient amounts of these antioxidants daily is expected to delay aging, although this effect on humans has not been elucidated.

Foods that have antiaging effects based on another mechanism are present. Rutin, a component of buckwheat and onion, improves blood circulation and insulin secretion [27, 28]. Sesamin protects many kinds of tissues. Allicin, contained in garlic and green onions, improves the secretion of several hormones and the reproductive function, and allithiamine promotes fatigue recovery [29]. Taurine, from octopus and shellfish, has effects of fatigue recovery and decrease of stress. Therefore, positive and frequent eating of buckwheat, sesame, garlic, and octopus may delay aging by controlling homeostasis. These foods are frequently used in the Japanese cuisine.

Moreover, important genes related to aging were discovered by the progress of recent studies [30]. One is the sirtuin gene, which restricts gene transcription and results in delay of aging. The sirtuin gene can be induced by caloric restriction. For instance, a group of researchers from Wisconsin University investigated the effects of caloric restriction on aging using rhesus monkeys; the aging effect was decreased in monkeys fed a low-calorie diet compared to the control group [31]. Other researchers suggested that resveratrol, component of red wine, is capable of enhancing the sirtuin gene [32]. The effect of polyamine on aging is also remarkable [33]. A previous study reported that the mice fed a high polyamine diet appeared much younger than the control. Another study reported that polyamine secreted by *Bifidobacteria* in yogurt is responsible for increasing longevity. Although the study on foods that have a capacity of rejuvenation is at a primary stage, such foods will become one of the targets as superfoods in the near future.

2.3. Foods to prevent cancer

Deaths caused by cancer are increasing worldwide, despite development of several technologies to discover the early stages of cancer. Among the various causes of cancer, dietary habits account for approximately 30%, while genetic factors have a less contribution. Therefore, positive and frequent eating of foods containing anti-cancer ingredients might be effective in preventing cancer.

The effectiveness of green tea [34], garlic [35], and soybean [36] in preventing cancer has been experimentally and statistically estimated. For instance, the rate of gastric cancer is much lower in Chinese and Italian people, who commonly eat garlic, and in Japanese people, due to the consumption of green tea. The rate of breast cancer in Oriental people, who commonly eat soybean, is lower than that in the Western people. These results suggest that allicin, daidzein (or genestin), and catechin are effective in preventing cancer.

It has also been reported that isocyanate, a component in cabbages, can prevent cancer by degrading nitrosamine compounds. DHA and oleic acids can inhibit the growth of cancer cells at the primary stage [37, 38]. Moreover, many kinds of seafood and mushrooms, which are frequently used in Japan, show anti-cancer effects. Fucoidan is a component of seaweeds, such as *Mozuku*, *Wakame* and *Hijiki*, and can inhibit both growth of cancer cells and formation of new blood vessels. β-Glucan contained in mushrooms, such as *Shitake* and *Maitake*, induces apoptosis of cancer cells by activating immune cells, for example, NK cells [39, 40].

Although many compounds have been discovered as anti-cancer agents, it is difficult to determine the most effective ingredient because various causative genes are also responsible for the disease. The National Cancer Institute in the United States conducted the "Designer Foods Project" in the 1990s. The purpose of this project was to categorize the importance of foods to cancer. As a result, approximately 40 foods with strong anticancer effect were selected, divided into three categories based on their importance, and arranged into a pyramidal shape. The foods belonging to the most important category are garlic, soybean, cabbage, ginger and plants belonging to the family Umbelliferae.

2.4. Foods to prevent dementia

The proportion of elderly people is rapidly increasing in Japan. Consequently, dementia, such as Alzheimer's disease (AD), is a serious social issue in Japan. Although studies have elucidated why AD is developed, effective treatment to cure dementia has not been established yet. Therefore, good dietary habits and lifestyle are the best ways to prevent the onset of AD.

DHA is one of key compounds effective in preventing AD. Phospholipids of the nerve cell membranes are rich in ω-3 PUFAs, and 11% of PUFAs are DHA, keeping a healthy network between nerve cells in hippocampus where new information is remembered by reconnecting a neural network. The conversion rates from α-linolenic acid to EPA and from EPA to DHA in humans are only 5 and 0.5%, respectively. Humans are unable to synthesize DHA in the brain. Therefore, the healthy function of the brain is strongly disturbed by the shortage of DHA. In fact, DHA content in the brain of AD patients is much lower than that of younger people [41]. Several studies have reported that the brain functions can be improved by frequent intake of DHA [42–45]. These results suggest that positive and frequent eating of fish, which contains DHA, is necessary to delay the onset of AD.

Inhibition of amyloid β ($A\beta$) accumulation and aggregation is also effective to prevent AD. Neprilysin (NEP) and insulin-degrading enzyme (IDE) function as $A\beta$ degrading enzymes [46, 47]. They preferentially function to degrade insulin, under high levels insulin, and, therefore, the incidence of AD is high in patients with diabetes whose insulin levels are constantly high [48]. Active ingredients, which can control the blood glucose level, are also effective against AD, that is, green tea [49].

In addition to the abovementioned compounds, several active ingredients are also effective. For instance, the effect of olive oil was discovered when the relationship between the diet and dementia was investigated in the Mediterranean region, where the rate of dementia is low. Some studies suggest that oleocanthal, from extra virgin olive oil, is the active ingredient effective in preventing AD [50, 51]. Dipropyl trisulfide (DPTS), contained in onions, is an antioxidant, whose preventive effect on dementia is expected because DPTS can be transferred to brain through the blood brain barrier. The rate of dementia in India is much lower than in America, probably due to consumption of curcumin, a component of turmeric [52]. Several studies have demonstrated the effectiveness of curcumin with regard to AD. Japanese frequently eat "curry rice" containing high amounts of turmeric.

3. Ideal diet using common foods containing active ingredients

Foods and active ingredients described in Section 2 were summarized in **Table 2**. These foods are commonly used in Japanese cuisines. As shown in **Table 2**, most of the active ingredients can prevent two or more diseases. A well-balanced diet containing these foods would be ideal to prevent lifestyle-related diseases, aging, cancer, and dementia. In Japanese food as well as in other cuisines, many kinds of fish, seaweeds, shellfish, and mushrooms are used. Therefore, well-balanced intake of active ingredients in a variety of dishes is a characteristic of Japanese food, and I believe that Japanese food is one of the ideal diets.

The Mediterranean diet is also noteworthy, since the risk factors of lifestyle-related diseases in people living in this region are low. Furthermore, many studies have elucidated that the Mediterranean diet is effective in preventing dementia as well as lifestyle-related diseases [53–55]. The Mediterranean diet is a good combination of fish, olive oil, shellfish, seafoods, and plant-based foods, such as vegetables, fruits and beans. Recently, a Mediterranean–DASH Intervention for Neurodegenerative Delay (MIND) diet was developed by a group of the Rush University [56–58]. The concept of the MIND is based on the combination of both advantages of the DASH and Mediterranean diets. In the MIND diet, positive intake of 10 kinds of foods and negative intake of 5 kinds of foods are recommended.

A remarkable aspect is that the active ingredients in Mediterranean diet and MIND diet are similar to those in Japanese food, and all of them utilize well-balanced intake of these active ingredients in common foods. Therefore, these active components are essential factors to realize an ideal diet. Foods and taste preferences are different for each country, and eating Japanese or Mediterranean foods every day can be difficult or unacceptable for people of

Active ingredients Contained in food	Effectiveness to A	B	C	D	Foods	Japanese cuisine
Unsaturated fatty acid						
DHA and EPA	○	○	○	○	Saury, mackerel, horse mackerel etc.	Sushi
Oleic acid	○	—	○	○	Olive oil	Anthocyanin
Antioxidants						
Anthocyanin	○	○	—	○	Berries	
Catechin	○	○	○	○	Green tea	Japanese tea
Daidzein and genestin	○	○	○	—	Soybean	Tofu, Miso
Rutin	○	○	—	○	Buckwheat	Soba
Lycopene	○	—	—	—	Tomato	
Resveratrol	—	○	—	—	Red wine	
Curcumin	—	—	○	○	Turmeric	Curry
Other compounds						
Allicin	○	○	○	—	Garlic, green onions	
Taurine	○	○	—	—	Octopus, turban shell	Takoyaki
Sesaminol	○	○	○	—	Sesame	
Alginate	○	—	—	—	Kelp, wakame	Miso soup
Fucoidan	—	—	○	—	Mozuku	
β-Glucan	—	—	○	—	Shiitake, maitake etc.	

A, Obesity and metabolic syndrome; B, aging; C, cancer; D, Alzheimer's disease; ○, effective; —, effectiveness is unknown or low effective.

Table 2. Active ingredients and foods commonly used in Japanese cuisines and their respective effectiveness to some diseases.

other countries. It is important to use local foods and cuisines for designing the ideal diet containing active ingredients.

4. Conclusion

Superfoods listed in **Table 1** have excellent effects on nutritional supplementation or on prevention diseases. Therefore, they are useful for the people who are short of medicine. Moreover, the study on the development of novel superfoods including genetically modified one is useful for food shortage, which may occur in the near future. On the other hand, several common foods that are frequently eaten daily contain sufficient amounts of active ingredients. These can be found in ideal diets, such as Japanese food, the Mediterranean diet, and the MIND diet. A well-balanced diet combination of these foods may be more effective in preventing diseases.

References

[1] Waisundara V, Shiomi N. Diabetes miletus in South Asia. In: Ahmed RG editors. Diabetes and Its Complications. Rijeca: InTech; 2018. p. 1-21. DOI: 10.5772/intechopen.76391

[2] Williamson C. Functional foods: What are the benefits? British Journal of Community Nursing. 2009;**14**(6):230-236. DOI: 10.12968/bjcn.2009.14.6.42588

[3] Pratt SG, MK. SuperFoods Rx: Fourteen Foods That Will Change Your Life. New York: Harper Collins; 2005

[4] Wolfe D. Superfoods: The Food and Medicine of the Future. California: North Atlantic Books; 2009

[5] Waisundara V, Shiomi N, editors. Superfood and Functional Foods – An Overview of Their Processing and Utilization. Rejeca: InTech; 2017. DOI: 10.5772/63180

[6] Shiomi N, Waisundara V, editors. Superfood and Functional Foods – The Development of Superfoods and their Roles as Medicine. Rejeca: InTech; 2017. DOI: 10.5772/65088

[7] Samson SL, Garber AJ. Metabolic syndrome. Endocrinology and Metabolism Clinics of North America. 2014;**43**(1):1-23. DOI: 10.1016/j.ecl.2013.09.009

[8] Chan JM, Rimm EB, Colditz GA, Stampfer MJ, Willett WC. Obesity, fat distribution, and weight gain as risk factors for clinical diabetes in men. Diabetes Care. 1994;**17**(9):961-969. DOI: 10.2337/diacare.17.9.961

[9] Huang CW, Chien YS, Chen YJ, Ajuwon KM, Mersmann HM, Ding ST. Role of n-3 polyunsaturated fatty acids in ameliorating the obesity-induced metabolic syndrome in animal models and humans. International Journal of Molecular Sciences. 2016;**17**(10). pii: E1689. DOI: 10.3390/ijms17101689

[10] Picklo MJ, Murphy EJ. A high-fat, high-oleic diet, but not a high-fat, saturated diet, reduces hepatic α-linolenic acid and eicosapentaenoic acid content in mice. Lipids. 2016;**51**(5): 537-547. DOI: 10.1007/s11745-015-4106-9

[11] Picklo Sr MJ, Idso J, Seeger DR, Aukema HM, Murphy EJ. Comparative effects of high oleic acid vs high mixed saturated fatty acid obesogenic diets upon PUFA metabolism in mice. Prostaglandins, Leukotrients and Essential Fatty Acids. 2017;**119**:25-37. DOI: 10.1016/j.plefa.2017.03.001

[12] Mitjavila MT, Fandos M, Salas-Salvadó J, et al. The mediterranean diet improves the systemic lipid and DNA oxidative damage in metabolic syndrome individuals. A randomized, controlled, trial. Clinical Nutrition. 2013;**32**(2):172-178. DOI: 10.1016/j.clnu.2012.08.002

[13] Murata T, Kuno T, Hozumi M, Tamai H, Takagi M, Kamiwaki T, ITOH Y. Inhibitory effect of calcium (derived from eggshell) -supplemented chocolate on absorption of fat in human males. Journal of Japan Society of Nutrition and Food Sciences (Japanese). 1998;**51**(4):165-174

[14] Fraga CG, Galleano M, Verstraeten SV, Oteiza PI. Basic biochemical mechanisms behind the health benefits of polyphenols. Molecular Aspects of Medicine. 2010;**31**(6):435-445. DOI: 10.1016/j.mam.2010.09.006

[15] Yang J, Kim CS, Tu TH, Kim MS, Goto T et al. Quercetin protects obesity-induced hypothalamic inflammation by reducing microglia-mediated inflammatory responses via HO-1 induction. Nutrients. 2017;**9**(7). pii: E650. DOI: 10.3390/nu9070650

[16] Singh BN, Shankar S, Srivastava RK. Green tea catechin, epigallocatechin-3-gallate (EGCG): Mechanisms, perspectives and clinical applications. Biochemical Pharmacology. 2011;**82**(12):1807-1821. DOI: 10.1016/j.bcp.2011.07.093

[17] LGeorg Jensen M, Kristensen M, Astrup A. Effect of alginate supplementation on weight loss in obese subjects completing a 12-wk energy-restricted diet: A randomized controlled trial. The American Journal of Clinical Nutrition. 2012;**96**(1):5-13. DOI: 10.3945/ajcn.111.025312

[18] Borck PC, Vettorazzi JF, Branco RCS et al. Taurine supplementation induces long-term beneficial effects on glucose homeostasis in ob/ob mice. Amino Acids. 2018 (in press). DOI: 10.1007/s00726-018-2553-3

[19] Shiomi N, Maeda M, Mimura M. Compounds that inhibit triglyceride accumulation and TNFα secretion in adipocytes. Journal of Biomedical Science and Engineering. 2011;**4**(11):684-691. DOI: 10.4236/jbise.2011.411085

[20] Kontani Y, Wang Y, Kimura K, Inokuma K, et al. UCP1 deficiency increases susceptibility to diet-induced obesity with age. Aging Cell. 2005;**4**(3):147-155. DOI: 10.1111/j.1474-9726.2005.00157.x

[21] Kozak LP, Anunciado-Koza R. UCP1: Its involvement and utility in obesity. International Journal of Obesity (London). 2008;**32**(Suppl 7):S32-S38. DOI: 10.1038/ijo.2008.236

[22] Shiomi N, Ito M, Watanabe K. Characteristics of beige adipocytes induced from white adipocytes by *Kikyo* extract. Journal of Biomedical Science and Engineering. 2016;**9**(7): 342-353. DOI: 10.4236/jbise.2016.97029

[23] Ono K, Tsukamoto-Yasui M, Hara-Kimura Y, Inoue N, Nogusa Y, Okabe Y, Nagashima K, Kato F. Intragastric administration of capsiate, a transient receptor potential channel agonist, triggers thermogenic sympathetic responses. Journal of Applied Physiology (1985). 2011;**110**(3):789-798. DOI: 10.1152/japplphysiol.00128.2010

[24] Liochev SI. Reactive oxygen species and the free radical theory of aging. Free Radical Biology and Medicine. 2013;**60**:1-4. DOI: 10.1016/j.freeradbiomed.2013.02.011

[25] Fusco D, Colloca G, Lo Monaco MR, Cesari M. Effects of antioxidant supplementation on the aging process. Clinical Interventions in Aging. 2007;**2**(3):377-387

[26] Sadowska-Bartosz I, Bartosz G. Effect of antioxidants supplementation on aging and longevity. BioMed Research International. 2014;**2014**:404680. DOI: 10.1155/2014/404680

[27] Li T, Chen S, Feng T, Dong J, Li Y, Li H. Rutin protects against aging-related metabolic dysfunction. Food and Function. 2016;**7**(2):1147-1154. DOI: 10.1039/c5fo01036e

[28] Jyrzanowska J, Piechal A, Blecharz-Klin K, Joniec-Maciejak I, Zobel A, Widy-Tyszkiewicz E. Influence of long-term administration of rutin on spatial memory as well as the concentration of brain neurotransmitters in aged rats. Pharmacological Reports. 2012;**54**:818-816

[29] Rahman MS. Allicin and other functional active components in garlic: Health benefits and bioavailability. International Journal of Food Properties. 2007;**10**:245-268. DOI: 10.1080/10942910601113327

[30] Shiomi N. Introductory chapter: Recent studies on cellular aging and rejuvenation. In: Shiomi N, editor. Molecular Mechanisms of the Aging Process and Rejuvenation. Rijeca: InTech; 2016. pp. 1-11. DOI: 10.5772/63897

[31] Mattison JA, Roth GS, Beasley TM, et al. Impact of caloric restriction on health and survival in rhesus monkeys from the NIA study. Nature. 2012;**489**(7415):318-321. DOI: 10.1038/nature11432

[32] Baur JA, Pearson KJ, Price NL, et al. Resveratrol improves health and survival of mice on a high-calorie diet. Nature. 2006;**444**(7117):337-342. DOI: 10.1038/nature05354

[33] Kibe R, Kurihara S, Sakai Y, Suzuki H, et al. Upregulation of colonic luminal polyamines produced by intestinal microbiota delays senescence in mice. Scientific Reports. 2014;**4**:4548. DOI: 10.1038/srep04548

[34] Fujiki H, Watanabe T, Sueoka E, Rawangkan A, Suganuma M. Cancer prevention with green tea and its principal constituent, EGCG: From early investigations to current focus on human cancer stem cells. Molecules and Cells. 2018;**41**(2):73-82. DOI:10.14348/molcells.2018.2227

[35] Luo R, Fang D, Hang H, Tang Z. The mechanism in gastric cancer chemoprevention by allicin. Anticancer Agents in Mediclinical Chemistry. 2016;**16**(7):802-809

[36] Sarkar FH, Li Y. Soy isoflavones and cancer prevention. Cancer Investigation. 2003;**21**(5): 744-757. DOI: 10.2174/1871520616666151111115443

[37] Newell M, Baker K, Postovit LM, Field CJ. A critical review on the effect of docosahexaenoic acid (DHA) on cancer cell cycle progression. International Journal of Molecular Sciences. 2017;**18**(8). pii: E1784. DOI: 10.3390/ijms18081784

[38] Boss A, Bishop KS, Marlow G, Barnett MP, Ferguson LR. Evidence to support the anti-cancer effect of olive leaf extract and future directions. Nutrients. 2016;**8**(8). pii: E513. DOI: 10.3390/nu8080513

[39] Atashrazm F, Lowenthal RM, Woods GM, Holloway AF, Dickinson JL. Fucoidan and cancer: A multifunctional molecule with anti-tumor potential. Marine Drugs. 2015;**13**(4): 2327-2346. DOI: 10.3390/md13042327

[40] Akramiene D, Kondrotas A, Didziapetriene J, Kevelaitis E. Effects of beta-glucans on the immune system. Medicina (Kaunas, Lithuania). 2007;**43**(8):597-606

[41] de Wilde MC, Vellas B, Girault E, Yavuz AC, Sijben JW. Lower brain and blood nutrient status in Alzheimer's disease: Results from meta-analyses. Alzheimers and Dement (New York, NY) 2017;**3**(3):416-431. DOI: 10.1016/j.trci.2017.06.002

[42] Hashimoto M. N-3 fatty acids and the maintenance of neuronal functions. Nihon Yakurigaku Zasshi. 2018;**151**(1):27-33. DOI: 10.1254/fpj.151.27

[43] Su HM. Mechanisms of n-3 fatty acid-mediated development and maintenance of learning memory performance. Journal of Nutritional Biochemistry. 2010;**21**(5):364-373. DOI: 10.1016/j.jnutbio.2009.11.003

[44] Newton W, McManus A. Consumption of fish and Alzheimer's disease. The Journal of Nutrition, Health and Aging. 2011 **15**(7):551-4552

[45] Wu S, Ding Y, Wu F, Li R, Hou J, Mao P. Omega-3 fatty acids intake and risks of dementia and Alzheimer's disease: A meta-analysis. Neuroscience and Biobehavioral Reviews. 2015;**48**:1-9. DOI: 10.1016/j.neubiorev.2014.11.008

[46] Wang S, Wang R, Chen L, Bennett DA, Dickson DW, Wang DS. Expression and functional profiling of neprilysin, insulin-degrading enzyme, and endothelin-converting enzyme in prospectively studied elderly and Alzheimer's brain. Journal of Neurochemistry. 2010;**115**(1):47-57. DOI: 10.1111/j.1471-4159.2010.06899.x

[47] Hüttenrauch M, Baches S, Gerth J, Bayer TA, Weggen S, Wirths O. Neprilysin deficiency alters the neuropathological and behavioral phenotype in the 5XFAD mouse model of Alzheimer's disease. Journal of Alzheimers Diseases. 2015;**44**(4):1291-1302. DOI: 10.3233/JAD-142463

[48] Schilling MA. Unraveling Alzheimer's: Making sense of the relationship between diabetes and Alzheimer's disease. Journal of Alzheimers Disease. 2016;**51**(4):961-977. DOI: 10.3233/JAD-150980

[49] Yasutake Tomata Y, Sugiyama K, Kaiho Y, et al. Green tea consumption and the risk of incident dementia in elderly Japanese: The Ohsaki cohort 2006 study. The American Journal of Geriatric Psychiatry. 2016;**24**(10):881-889. DOI: 10.1016/j.jagp.2016.07.009

[50] Rigacci S. Olive oil phenols as promising multi-targeting agents against Alzheimer's disease. Advances in Experimental Medicine and Biology. 2015;**863**:1-20. DOI: 10.1007/978-3-319-18365-7_1

[51] Qosa H, Mohamed LA, Batarseh YS, Alqahtani S, Ibrahim B, LeVine 3rd H, Keller JN, Kaddoumi A. Extra-virgin olive oil attenuates amyloid-β and tau pathologies in the brains of TgSwDI mice. Journal of Nutritional Biochemistry. 2015;**26**(12):1479-1490. DOI: 10.1016/j.jnutbio.2015.07.022

[52] Mishra S, Palanivelu K. The effect of curcumin (turmeric) on Alzheimer's disease: An overview. Annals of Indian Academy of Neurology. 2008;**11**(1):13-19. DOI: 10.4103/0972-2327.40220

[53] Sanches Machado d'Almeida K, Ronchi Spillere S, Zuchinali P, Corrêa Souza G. Mediterranean diet and other dietary patterns in primary prevention of heart failure and changes in cardiac function markers: A systematic review. Nutrients. 2018;**10**(1). pii: E58. DOI: 10.3390/nu10010058

[54] Salas-Salvadó J, Guasch-Ferré M, Lee CH, Estruch R, Clish CB, Ros E. Protective effects of the mediterranean diet on type 2 diabetes and metabolic syndrome. Journal of Nutrition. 2016;**146**(4):920S-927S. DOI: 10.3945/jn.115.218487

[55] Radd-Vagenas S, Duffy SL, Naismith SL, Brew BJ, Flood VM, Fiatarone Singh MA. Effect of the mediterranean diet on cognition and brain morphology and function: A systematic review of randomized controlled trials. American Journal of Clinical Nutrition. 2018;**107**(3):389-404. DOI: 10.1093/ajcn/nqx070

[56] Morris MC, Tangney CC, Wang Y, Sacks FM, Barnes LL, Bennett DA, Aggarwal NT. MIND diet slows cognitive decline with aging. Alzheimer's and Dementia 2015;**11**(9):1015-1022. DOI: 10.1016/j.jalz.2015.04.011. (Epub Jun 15, 2015)

[57] Morris MC, Tangney CC, Wang Y, Sacks FM, Bennett DA, Aggarwal NT. MIND diet associated with reduced incidence of Alzheimer's disease. Alzheimer's and Dementia. 2015;**11**(9):1007-1014. DOI: 10.1016/j.jalz.2014.11.009

[58] Morris MC. Nutrition and risk of dementia: Overview and methodological issues. Annals of New York Academy of Sciences. 2016;**1367**(1):31-37. DOI: 10.1111/nyas.13047

Chapter 2

Bioactive Compounds Contained in Mediterranean Diet and Their Effects on Neurodegenerative Diseases

Abstract

Neuroinflammatory processes in the brain are believed to play a crucial role in the development of neurodegenerative diseases, especially due to increased production of reactive oxygen species. The brain is susceptible to oxidative stress more than other organs due to the low activity of antioxidant defense systems. In agreement with these observations, increased oxidative stress plays an important role in the pathogenesis of neurodegenerative diseases such as Alzheimer disease, Parkinson disease, ischemic diseases and aging. The Mediterranean diet is inspired by the traditional dietary pattern of some countries of the Mediterranean basin. From ancient times, these populations were characterized by simple food habits as high intake of whole cereals (pasta, bread, rice), fruits and vegetables (up to 400 g day^{-1} in Greece), legumes and fish, olive oil as the common source of fats, poor intake of meat and dairy products and a moderate, regular wine drinking. In the present chapter, there are going to be presented some bioactive substances present in the Mediterranean diet related to the prevention of neurodegenerative diseases. These substances are able to exert important antioxidant activity (through mechanisms such as sequestration of free radicals, inhibition of the production of hydrogen peroxide, activation of endogenous defense mechanisms.

Keywords: Mediterranean diet, neurodegenerative diseases, Alzheimer disease, Parkinson disease, melatonin, hydroxytyrosol

1. Introduction

Neurodegeneration in Parkinson's, Alzheimer disease, and other neurodegenerative diseases seems to be multifactorial, in that a complex set of toxic reactions leads to the demise of neurons. Complications include: inflammation, glutamatergic neurotoxicity, increases in iron and nitric oxide, depletion of endogenous antioxidants, reduced expression of trophic factors, dysfunction of the ubiquitin-proteasome system, and expression of proapoptotic proteins leads to the demise of neurons [1].

At pathological level, almost all neurodegenerative diseases share common features such as the generation of misfolded protein deposits, metal ion deregulation, and exposure to oxidative stress [2, 3]. In Alzheimer disease the extracellular senile plaques are consisted on amyloid-β peptides derived from the mutations in genes encoding the amyloid precursor protein while the intracellular tangles are from hyperphosphorylated Tau protein. In Parkinson disease the accumulation of intracytoplasmic Lewy bodies is mainly composed of α-synuclein and ubiquitin [4].

Alzheimer's disease is responsible of 70% cases of dementia in elderly people. Between 2000 and 2013, United States of America death rates for dementia increased 21% for men and 31% for women. Among individuals 85 age old or older, dementia-associated death rates for women and men were ~4% and 3.2%, respectively [5]. Risk factors include hypercholesterolemia, obesity, diabetes, and cardiovascular factors, such as hypertension, and inflammation [6].

Nowadays, neurodegenerative diseases are not curable and treatments have limited effectiveness. Hence, increasing interest for effective preventive measures has been recently shown in the scientific literature [7]. Prevention of neurodegenerative diseases and search for new drugs are the great challenges of scientific research, because the symptoms appear in the human being only when the degeneration is advanced. Mechanisms involved in neurodegenerative diseases are complex and multifactorial. However, these mechanisms present common pathways, including: mitochondrial dysfunction, intracellular Ca^{2+} overload, oxidative stress and inflammation. Often multiple pathways coexist, restricting benefits from therapeutic interventions.

Neuroinflammatory processes in the brain are believed to play a crucial role in the development of neurodegenerative diseases, especially due to increased production of reactive oxygen species. The brain is susceptible to oxidative stress more than other organs due to the low activity of antioxidant defense systems. In agreement with these observations, increased oxidative stress plays an important role in the pathogenesis of neurodegenerative diseases such as Alzheimer disease, Parkinson disease, ischemic diseases and aging [8].

Several models of diet have been proposed but, until now, the highest attention of researchers, clinicians, and institutions has been focused on the Mediterranean diet. This diet has been promoted as a model for healthy eating and it has been widely recognized to have favorable effects on lipid profile and to provide a significant source of antioxidants and vitamins [9].

The Mediterranean diet is inspired by the traditional dietary pattern of some countries of the Mediterranean basin. From ancient times, these populations were characterized by simple food habits as high intake of whole cereals (pasta, bread, rice), fruits and vegetables (up to 400 g day^{-1} in Greece), legumes and fish, olive oil as the common source of fats, poor intake of meat and dairy products and a moderate, regular wine drinking. The intake of saturated animal fats is relatively low, and moderate fish consumption gives enough polyunsaturated fatty acids [9, 10].

Valls-Pedret et al. and Ngandu et al. provide a strong level of scientific evidence for the beneficial effects of the Mediterranean diet on cognitive functions [11, 12]. Several clinical, epidemiological and experimental studies suggest that consumption of the Mediterranean diet reduces the incidence of certain pathologies related to oxidative stress, chronic inflammation and immune system diseases such as cancer, atherosclerosis, cardiovascular disease and neurodegenerative diseases [13]. These reductions can be partially attributed to different bioactive compounds present in the Mediterranean diet (omega 3 fatty acids, polyphenols, resveratrol or melatonin). In fact, the five most important adaptations induced by the Mediterranean dietary pattern are [7]:

1. Lipid lowering effect,
2. Protection against oxidative stress, inflammation and platelet aggregation,
3. Modification of hormones and growth factors involved in the pathogenesis of cancer,
4. Inhibition of nutrient sensing pathways by specific amino acid restriction, and
5. Gut microbiota-mediated production of metabolites influencing metabolic health.

There is negative correlation between cognitive functions, saturated fatty acids and protective effect against cognitive decline with increased fish consumption, high intake of monounsaturated fatty acids and polyunsaturated fatty acids (PUFA), particularly n-3 PUFA [14]. Similarly, polyunsaturated and omega-3 fatty acids as docosahexaenoic acid (DHA) and eicosapentaenoic acid (EPA) supplements are associated with increased cognitive function, due to the cumulating of factors that ultimately favor membrane permeability and neuronal functioning [15].

There are numerous epidemiological studies relating the Mediterranean diet with the prevention of neurodegenerative diseases and minor cognitive decline [14, 16]. However, other studies did not observe such relationship. Anastasiou et al. suggest that adherence to the Mediterranean diet is associated with better cognitive performance and reduced dementia in Greek elderly population [17]. Hardman et al. demonstrate that the relationship between Mediterranean diet score and cognition was only significant when medication use was taken into account.

In the present chapter, there are going to be presented some bioactive substances present in the Mediterranean diet related to the prevention of neurodegenerative diseases. These substances are able to exert important antioxidant activity (through mechanisms such as sequestration of free radicals, inhibition of the production of hydrogen peroxide, activation of endogenous defense mechanisms (catalase, superoxide dismutase), chelation of metals, etc.). However, many other biologically plausible mechanisms may be responsible for their protective effect [18, 19] as follows:

- Modulation of gene expression.
- Detoxification of carcinogens.
- Induction of cell death.
- Protection of DNA.
- Modification of cellular communication.
- Modification of the hormonal profile.
- Modulation of the lipid profile.
- Stimulation of the immune system.

- Anti-inflammatory effect.
- Effects on hemostasis.
- Hypocholesterolemic effect.
- Hypotensive effect.
- Antimicrobial activity

2. Bioactive substances in the Mediterranean diet

2.1. Resveratrol

Resveratrol belongs to the family of stilbenes and is one of the most studied polyphenols, mostly present in grapes and wines. This stilbene was discovered in 1940 in *Veratrum grandiflorum* by Takaoka [20] and reported in high concentration in *Vitis vinifera* in 1976 by Langcake and Pryce [21], leading to the subsequent research about bioactive function of the molecule (**Figure 1**). The scientific literature has reported several benefits related to resveratrol, most of them regarding to antioxidant capacity and cardiovascular improvement. However, pharmacokinetics of the molecule and the great content on red wine and grapes reveal resveratrol as an essential compound on the Mediterranean diet regarding neuroprotection.

Despite several studies regarding pharmacokinetics of resveratrol, it is still uncertain. Many scientific researchers have reported excellent values regarding *in vitro* bioavailability of resveratrol [22]. However, these values are not in agreement with *in vivo* results, mainly due to chemical insubstantiality [23]. In fact, revisions about pharmacokinetics differ widely between

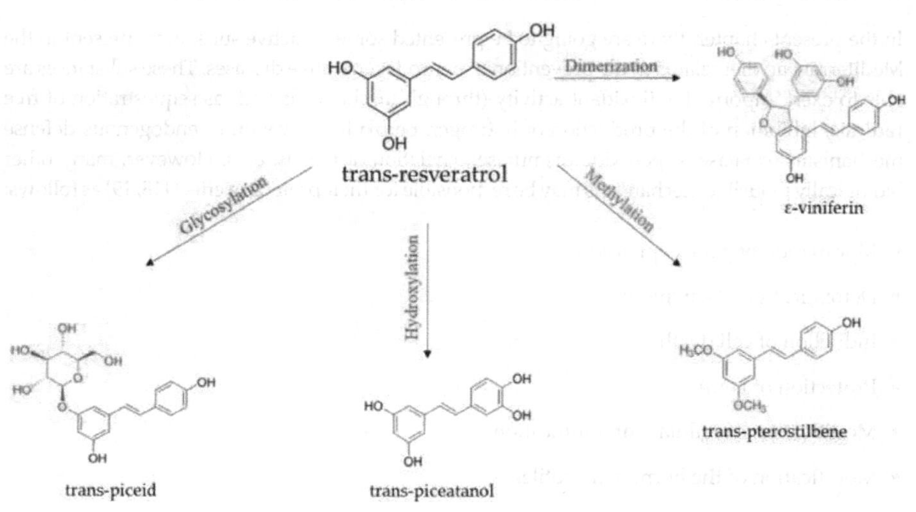

trans-resveratrol
Dimerization
ε-viniferin
Glycosylation
Hydroxylation
Methylation
trans-piceid
trans-piceatanol
trans-pterostilbene

Figure 1. Resveratrol and related metabolites.

them about bioavailability of resveratrol. Fernández-Mar et al. reviewed knowledge on scientific literature, reporting a bioavailability of resveratrol over 70% [24]. Meanwhile, Ahmed et al. found a bioavailability minor than 1% after extensive revision of literature, reflecting the high discrepancies about the absorption of resveratrol [23].

Dietary sources of resveratrol comprise a wide variety of plant matrices [23]. Health benefits attributed to resveratrol were noticed by the scientific community, leading to the increment of scientific reports relating its presence in a wide range of plants. From now, resveratrol has been reported to be present in at least 72 plant derived foods [23, 25]. As reported by Fernández-Mar et al. main sources of resveratrol comprise peanuts, pistachios, different berries, dark cocoa, and grapes/wine. All of them could be included in the Mediterranean diet; however, the main source of resveratrol in the Mediterranean diet are grape and derivatives, especially red wine [26–29]. Thus, concentration of trans-resveratrol in red wine can reach 14.3 mg/L, depending on the type of grape, cultivar conditions, or vinification procedures [23, 26–29]. Resveratrol is exclusively present in seeds and skin of grapes. Therefore, concentration of resveratrol in red wine is higher than white ones, due to the vinification process which leads to more extensive contact between skin/seeds and must [24]. As commented above, not only resveratrol has been found in grapes and must.

As reviewed by Ahmed et al. [23], resveratrol can also be found in different berries and related products, namely blueberries cranberries, bilberries, lingonberries, partridgeberries, mulberries and strawberries [30–32]. However, the different technological and agronomical processes applied to berries and the slight content on resveratrol, are limiting factors for considering berries as remarkable source of that bioactive compound. Similarly, other foods containing minor amounts of resveratrol are dark cocoa, beer, and bee wax from honeycomb [33–39].

Other food with high amount of resveratrol are peanuts, in which resveratrol can be found in all edible parts of the plant. For example, peanut butter and peanut oil were found to have high quantity of trans-resveratrol reaching 16.9 μg/g in case of peanut oil [36, 37].

Resveratrol has shown different pathways that could postpone neurodegenerative diseases onset, especially Alzheimer disease [23]. In 2017, Ahmed et al. published a revision about the role of resveratrol in Alzheimer disease and other neurodegenerative diseases. They showed a total of 18 recent publications about the different action of resveratrol in human organism [23].

As a summary, resveratrol leads to:

- Inhibition of tauopathy.
- Enhancement of long-term memory formation.
- Inhibition of brain pro-inflammatory responses.
- Inactivation of astrocytes.
- Enhancement of sirtuin-1 activity.
- Protection from oxidative injury.
- Inhibition of neurotoxicity by H_2O_2.

- Inhibition of synthesis of Aβ plaque.
- Prevention of neuronal death.

Cognitive impairment is also susceptible to be treated with resveratrol, as observed in animal models. Consequently, resveratrol might also be considered as a potential anti-depressant bioactive compound [28].

Neuronal deficiency originates from multiple neurodegenerative diseases. Resveratrol seems to be quite associated to the development of specific neurodegenerative diseases as Alzheimer disease or Huntington disease and Parkinson disease [29]. The most outstanding capacity of resveratrol that influences its effectiveness for the treatment of neurodegenerative diseases is the ability to penetrate the blood-brain barrier. In fact, it has demonstrated a great neuroprotective capacity, even in administration at low doses.

The activation of sirtuin-1 pathway (SIRT1) seems to be a determinant property of resveratrol [40, 41]. Parker et al. [42] showed that by daily intake of one wine glass, it brings enough resveratrol (500 nM) to combat neuronal dysfunction caused in Huntington and Alzheimer's diseases through SIRT1 activation.

Regarding Parkinson's disease, the first advances regarding its relation with resveratrol were published in the year 2000 [43]. In mice, Karuppagounder et al. reported that daily intake of resveratrol decreased Aβ plaque in the CNS. Mayor changes were observed in the medial cortex, the striatum and the hypothalamus. Moreover, the most noticeable changes were observed deprived of the sirtuin-1 pathway, which enhances the hypothesis of reduced formation of Aβ plaque due to reduction cysteine and glutathione in the CNS [44].

One of the most studied characteristics of resveratrol is its ability to reduce oxidative injury. Reactive oxygen species (ROS) are mayor agents promoted unpaired oxidative stress, and are determinant for the production of oxidative injury and non-enzymatic lipid peroxidation. Oxidized lipoproteins stimulates apoptosis due to the union of DNA to NF-κB. Resveratrol acts inhibiting the activation of NF-κB, which reduces the possibility of oxidative injury at the Central Nervous System (CNS) [45, 46].

2.2. Melatonin

Melatonin (*N*-acetyl-5-methoxytyramine) is a characteristic neurohormone of the pineal gland, also produced as a secondary metabolite in plants (**Figure 2**). It has been shown that synthesis of melatonin is produced from tryptophan, serotonin and *N*-acetylserotonin ultimately. On the other hand, melatonin molecule can also be formed by *O*-methylation of serotonin followed by *N*-acetylation of 5-methoxytryptamine in yeast [47, 48].

The absorption of melatonin after oral intake has been previously approached by the scientific literature, reporting similar values between different researchers. Bioavailability reported values vary from 33 to 8.7% [49–51]. The absorption of melatonin is delimited by many parameters such as age, sex, season or circadian cycle. Nevertheless, there is variables such as elimination and distribution half-life that look to remains equivalent between subjects.

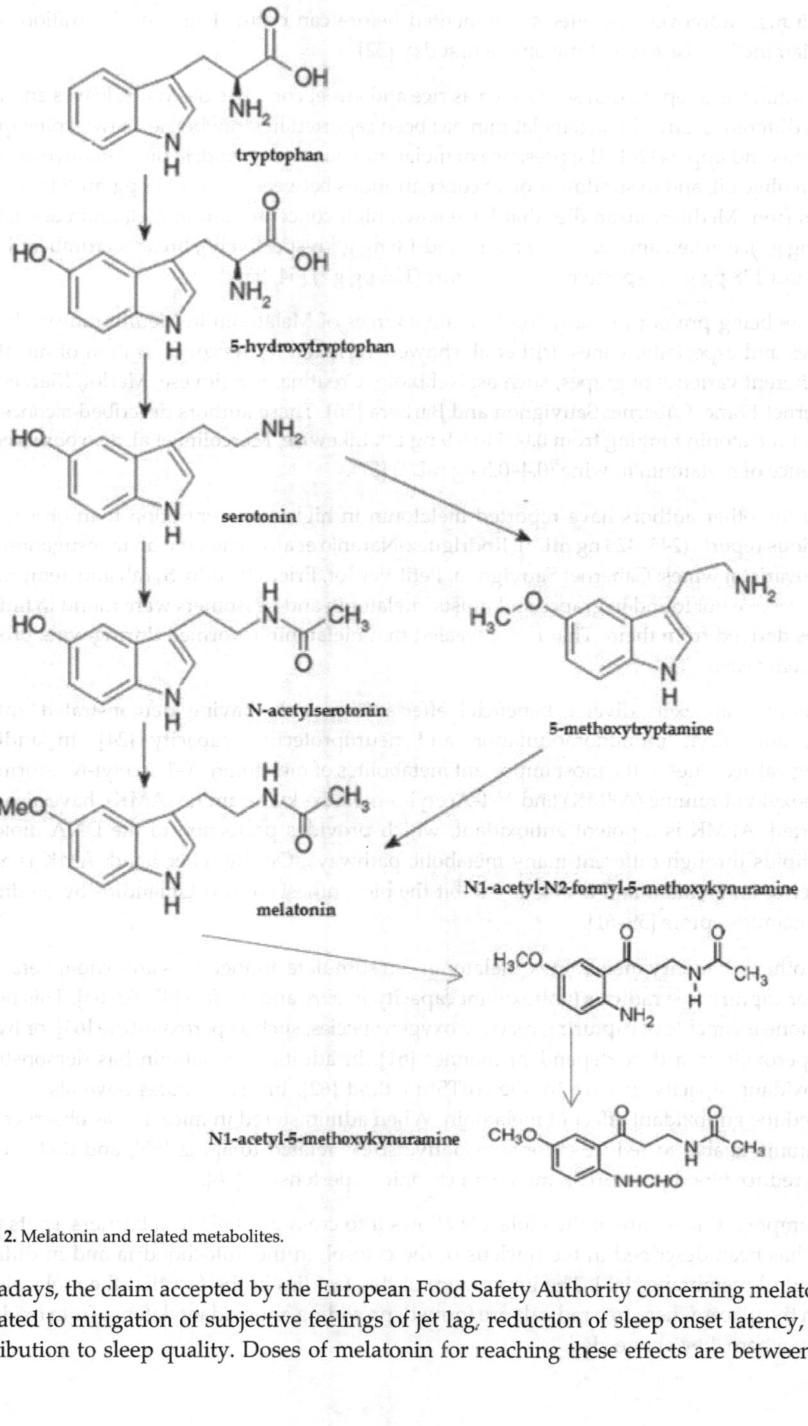

Figure 2. Melatonin and related metabolites.

Nowadays, the claim accepted by the European Food Safety Authority concerning melatonin is related to mitigation of subjective feelings of jet lag, reduction of sleep onset latency, and contribution to sleep quality. Doses of melatonin for reaching these effects are between 0.5

and 5 mg. Moreover, the effects commented before can occur if the administration of the indolamine is close to bed time on the first day [52].

Melatonin were reported in seeds such as rice and sweet corn, besides roots, leaves and fruits from different plants. In fact, melatonin has been reported in strawberries, kiwis, pineapples, bananas and apples [24]. The presence of melatonin has also been described in olive oil, extra virgin olive oil, and in sunflower oil at concentrations between 71 and 113 pg mL^{-1} [53]. Other foods from Mediterranean diet that have sown high concentration of melatonin are salmon (3.7 ng g^{-1}), chicken and lamb (2.3 ng g^{-1} and 1.6 ng g^{-1} respectively), bread's crumb and crust (341 and 138 pg g^{-1} respectively) and yogurt (126 pg g^{-1}) [54, 55].

Despite being present in many foods, main sources of Melatonin in Mediterranean diet are grapes and especially, wines. Iriti et al. showed variation in the concentration of melatonin in different varieties of grapes, such as: Nebbiolo, Croatina, Sangiovese, Merlot, Marzemino, Cabernet Franc, Cabernet Sauvignon and Barbera [56]. These authors described a concentration of melatonin ranging from 0.005 to 0.9 ng g^{-1}. Likewise, Mercolini et al. also observed the presence of melatonin in wine (0.4–0.5 ng mL^{-1}) [57].

Recently, other authors have reported melatonin in higher concentration than observed in previous reports (245–423 ng mL^{-1}). Rodríguez-Naranjo et al. carried out an investigation in 10 monovarietal wines: Cabernet Sauvignon, Petit Verdot, Prieto Picudo, Syrah and Tempranillo [58]. Despite not found in grapes and musts, melatonin and its isomers were found in finished wines derived from them. That fact, revealed that melatonin is formed during wine processing from yeasts.

Melatonin can exert diverse beneficial effects for health, having demonstrated antioxidant, anticancer, immunomodulation and neuroprotective capacity [24]. In addition, biological activities of the most important metabolites of melatonin, *N*-1-Acetyl-*N*-2-formyl-5-methoxykynuramine (AFMK) and *N*-1-Acetyl-5-methoxykynuramine (AMK), have also been reported. AFMK is a potent antioxidant, which provides protection to the DNA molecule and lipids through different many metabolic pathways. On the other hand, AMK is also a powerful antioxidant and is able to inhibit the biosynthesis of prostaglandins by binding to diazepam receptors [59–61].

Like other secondary metabolites, melatonin can stimulate endogenous antioxidant enzymes and/or capture free radicals (antioxidant capacity *in vitro* and *in vivo*) [47, 62, 63]. This neurohormone is capable of capturing reactive oxygen species, such as peroxynitrite [64]; or hydrogen peroxide in a dose-dependent manner [61]. In addition, melatonin has demonstrated antioxidant capacity in vitro by the $ABTS^{+}$ method [62]. In vivo studies have also demonstrated the antioxidant effect of melatonin. When administered in mice, it was observed that melatonin is able to reduce chronic oxidative stress related to aging [65], and that it could even reduce blood pressure in men with chronic hypertension [66].

The amphipathic nature of the molecule allows it to cross physiological barriers, so its presence has been described in the nucleus of the cytosol, in the mitochondria and in different biological membranes [67]. The importance of this fact lies in the fact that the molecule can act in the places where free radicals are formed, providing antioxidant defense from oxidative injury where they are needed.

The role of melatonin as neuroprotective agent is relevant. It has been successfully proved in sleep disorders, helping to restore circadian rhythm. Moreover, melatonin is especially effective in patients with neurodegenerative diseases [68]. Several scientific studies have been carried out with the aim for palliating consequences of diseases such as Alzheimer, Parkinson, Huntington disease or amyotrophic lateral sclerosis, obtaining satisfactory results [24].

Miller et al. widely reviewed neuroprotective capacity of melatonin, reporting a wide range of actions in the human being [69]. Regarding neurodegenerative diseases, the most susceptible to be treated by melatonin are Parkinson disease [70–72], Multiple sclerosis [73, 74], Alzheimer disease [75–77] and amyotrophic lateral sclerosis [78–81].

The different effects of melatonin that could be useful for the improvement of neurological pathologies are large [69]. Melatonin can improve the evolution of Parkinson disease by the reduction of excitotoxicity caused by the autoxidation of dopamine [70], or improving quality and length of sleep [72]. Moreover, melatonin protects from injury to mitochondria, decreasing lipid peroxidation in multiple sclerosis and Alzheimer disease [73–75] and increase antioxidant enzymes generation [74]. Melatonin has also showed capacity to protect against cognitive deficits and inhibits formation of nicotinamide and Aβ plaque [76–78]. Melatonin also is able to reduce oxidative injury by reducing carbonyls formation [79, 80], and delays the progression of amyotrophic lateral sclerosis by inhibiting MT1 reception loss [81].

2.3. Hydroxytyrosol

Hydroxytyrosol is also known as 2-(3,4-dihydroxyphenyl)-ethanol (3,4-DHPEA) and as DOPET (**Figure 3**). Hydroxytyrosol is mainly found in olive oil as secoiridoid derivatives, as acetate and in free form [82]. Both hydroxytyrosol and its derivatives arise from oleuropein (hydroxytyrosol esterified with elenolic acid), present in olives during the extraction of olive oil [24].

Wine has proven to be another important source of hydroxytyrosol in the Mediterranean diet, and is formed in wine from tyrosol during alcoholic fermentation. Hydroxytyrosol was firstly found in Italian wines by Di Tommaso et al. [83], and later in other Italian and Greek wines [84–86]. Some authors describe a higher concentration in red wines (3.66–4.20 mg L^{-1}) than in white wines (1.72–1.92 mg L^{-1}) [24, 87]. Finally, Minuti et al. obtained hydroxytyrosol concentrations between 1.8 and 3.1 mg L^{-1} in red wine [87]. Thus, scientific literature shows that wine is an important source of hydroxytyrosol in the diet, along with olive oil [24].

De La Torre et al. and Schröder et al. investigated the bioavailability of hydroxytyrosol by comparing the intake of red wine and olive oil [87–89]. The intake of red wine (250 mL) increased plasmatic concentration of hydroxytyrosol above 8 ng mL^{-1}, representing greater increase than the observed after the intake of olive oil (0.35 mg of hydroxytyrosol with the intake of wine red and 1.7 mg with the ingestion of olive oil). These authors proposed the endogenous production of hydroxytyrosol from ethanol and dopamine in response to the observed increase.

Moreover, tyramine has been proposed as another route for hydroxytyrosol formation, and could be partly responsible for the increased endogenous formation of hydroxytyrosol after the intake of red wine. Therefore, substantial content of tyramine in red wine, could lead to the increase of endogenous formation of hydroxytyrosol. However, the amount of tyramine in the red wine is not large enough to explain such increase in endogenous hydroxytyrosol [88].

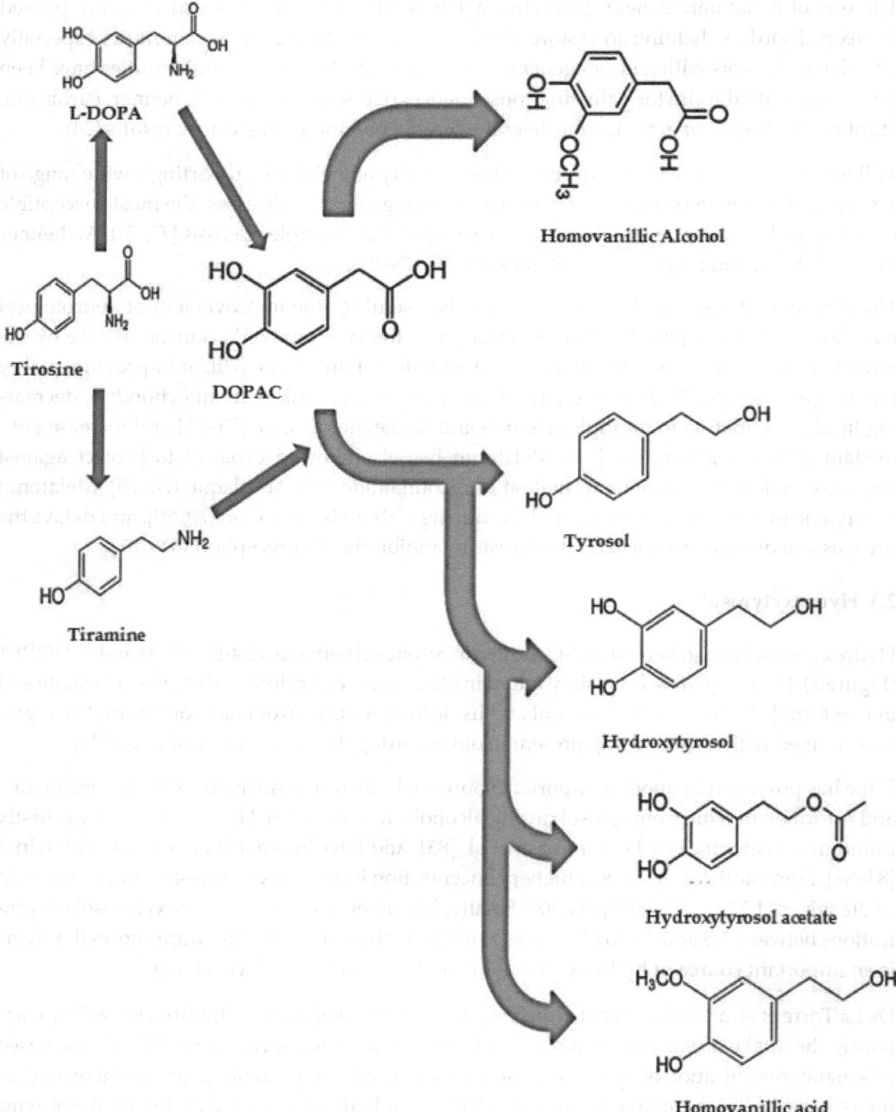

Figure 3. Hydroxytyrosol and related metabolites.

Several studies have reported beneficial effects of hydroxytyrosol, mainly after the intake of olive oil (the main source of hydroxytyrosol in the Mediterranean diet). The different effects attributed to hydroxytyrosol include: antioxidant capacity, cardioprotective effect, anticancer, antimicrobial, neuroprotective and antidiabetic activity [24]. Numerous authors have proved their ability both to chelate oxidizing compounds [90, 91], and to increase the concentration of antioxidant enzymes [91].

Neurodegenerative diseases, such as Alzheimer's disease or Parkinson's disease, could also be improved by the ingestion of hydroxytyrosol [92, 93]. As melatonin, hydroxytyrosol has the ability to cross the blood–brain barrier. Therefore, it can go through the brain and is rapidly metabolized, acting where the oxidative attack is produced [24].

Different studies evaluating the effects of hydroxytyrosol, showed the great neuroprotective capacity of the molecule. Hydroxytyrosol considerably inhibits LDL efflux in a dose-dependant way, both *in vivo* and *in vitro*. That fact offers an initial knowledge for more studies as potential effects of hydroxytyrosol as neuroprotective compound [93]. Marhuenda et al. reported a descend on the formation of neuroprostanes and F_2-dihomo-isoprostanes after the intake of red wine [28]. These effect was related to the content on hydroxytyrosol, more than other compounds from red wine matrix. Moreover, an oleuropein-enriched extract showed neuroprotective capacity by establishing a non-covalent complex with the amyloid-β-peptide, so can be decisive in many neurodegenerative diseases as Alzheimer's or Parkinson's disease. Therefore, hydroxytyrosol, being the main degradation molecule from oleuropein, can be proposed as a promising neuroprotective compound [92].

2.4. Polyphenols

Polyphenols comprises a large and heterogeneous group of phytochemicals containing phenol rings. They are mainly divided into flavonoids, phenolic acids, stilbenes, and lignans. Mayor flavonoids are flavones, flavonols, flavanols, flavanones, isoflavones, and anthocyanins [94] (**Table 1**).

Polyphenols can induce antioxidant enzymes such as glutathione peroxidase, catalase and superoxide dismutase that decompose hydroperoxides, hydrogen peroxide and superoxide anions, respectively. Moreover, they can also inhibit the expression of enzymes such as xanthine oxidase [95]. Dietary polyphenols have been shown to play important roles in human health. In fact, high intake of fruits, vegetables and whole grains, which are rich in polyphenols, has been related to reduced risk of many chronic diseases including cancer, cardiovascular disease, chronic inflammation and many degenerative diseases [96].

Health benefits of catechins and proanthocyanidins are related to their antioxidant character and free radical scavenger activity. Moreover, they can switch mechanisms involved in different pathologies such as: hypertension, inflammation, proliferation cellular, thrombogenesis, hypertriglycemia, hypercholesterolemia and neurodegenerative diseases or neuroinflammation [94].

Studies conducted on cell cultures stimulated with lipopolysaccharide show that the administration of quercetin, catechin and epigallocatechin gallate blocks the inflammatory response by inhibiting NOSi and the expression of cyclooxygenase (COX-2), as well as the production of NO, the release of pro-inflammatory cytokines, and the generation of ROS, in astrocytes and in microglia [97]. These extracts of phytochemicals have also been shown to inhibit MAPKs such

		Main food sources
Flavonoids	Anthocyanins	Cherries, red wine, olives, hazelnuts, almonds, black elderberry, black chokeberry, blueberries
	Flavonols	Grapefruit/pomelo juice, oranges, orange juice, grapefruit juice
	Flavanols	Dark chocolate red wine, apples, peaches, cocoa powder, nuts, dark chocolate
	Isoflavones	Soy flour, beans soy paste, roasted soy bean,
	Flavanones	Grapefruit/pomelo juice, oranges, orange juice, grapefruit juice
	Flavones	Virgin olive oil, oranges, whole grain wheat-flour bread, refined-grain wheat-flour bread, whole grain wheat four, black olives
Stilbenes	Resveratrol	Grapes, red wine, nuts
Phenolic acids	Benzoic acid	Olives, virgin olive oil, red wine, walnuts, pomegranate juice, red raspberry
		Coffee, maize oil, potatoes
Lignans	Cinnamic acid	Virgin olive oil, whole grain rye flour, bread from whole grain rye flour, flaxseed

Table 1. Main food sources of polyphenols [101].

as p38 or ERK1/2, which regulate NOSi and TNF-α, in addition to the activation of glial cells [97]. Mendel et al. suggest that catechins, may protect brain from aging and reduce the incidence of dementia, Alzheimer disease and Parkinson disease [98]. Moreover, Geiser et al. indicate that both anti-aggregation and antioxidant characteristics of catechins may alter mRNA expression to reduce feed-forward mechanisms and promote non-amyloidogenic processing [99].

In vivo studies show that chronic administration of epicatechin in combination with physical exercise, improves spatial memory, due to the increase in the Akt protein that activates the endothelial nitric oxide synthase (NOSe) enzyme, stimulate the angiogenesis, as well as the increase in neuronal density in regions such as the dentate gyrus of the hippocampus [100].

Other group of polyphenols which has showed several benefits are anthocyanins. Several studies have shown beneficial effects of anthocyanins on health, and the high antioxidant capacity due to their capacity to protect from free radicals by the donation of hydrogen atoms. The role of anthocyanins in neurodegenerative diseases is strongly linked to oxidative attack protection. Anthocyanins can modulate cognitive and motor function, enhancing memory, and preventing age-related decline in neural function [102]. Extracts rich in anthocyanins and proanthocyanidins exhibited greater neuroprotective activity than extracts rich in other polyphenols. Moreover, many individual anthocyanins interfered with rotenone neurotoxicity, which can be related with increased memory [103].

Finally, mayor food containing anthocyanins are berries that can effectively reverse age-related deficits in certain aspects of working memory. Anthocyanins and other flavonoids can prevent neuroinflammation, by the activation of synaptic signaling, and improving blood flow to the brain. It appears that some dietary anthocyanins can cross the blood–brain barrier, allowing the compounds to have a direct beneficial effect [100]. Anthocyanins suppress mitochondrial oxidative stress-induced apoptosis by preserving mitochondrial GSH and inhibiting cardiolipin oxidation and mitochondrial fragmentation [104].

3. Conclusion

Neurodegenerative diseases are a public health problem and the possibilities of delaying their evolution constitute a challenge for research. Considering the relationship between oxidative stress and neuroinflammation with neurodegenerative diseases, monitoring a diet rich in bioactive substances with antioxidant activity and polyunsaturated fatty acids of the omega-3 series, with a proven antiinflammatory and neuroprotective effect, could slow down the evolution of the disease, improve cognitive deterioration, delay the decline of motor symptoms and improve the quality of life of patients. An example of this type of diet is the Mediterranean diet, which is characterized by the consumption of fruits, vegetables, legumes, nuts, olive oil, moderate consumption of red wine and blue fish that incorporate bioactive substances with beneficial effects on health.

References

[1] Mandel S, Youdim MB. Catechin polyphenols: Neurodegeneration and neuroprotection in neurodegenerative diseases. Free Radical Biology and Medicine. 2004;**37**(3):304-317

[2] Roberts BR, Ryan TM, Bush AI, Masters CL, Duce JA. The role of metallobiology and amyloid-β peptides in Alzheimer's disease. Journal of Neurochemistry. 2012;**120**(s1): 149-166

[3] Granzotto A, Zatta P. Resveratrol and Alzheimer's disease: Message in a bottle on red wine and cognition. Frontiers in Aging Neuroscience. 2014;**6**:9

[4] La Spada AR. Finding a sirtuin truth in Huntington's disease. Nature Medicine. 2012; **18**(1):24-26

[5] Selmin OI, Romagnolo AP, Romagnolo DF. Mediterranean diet and neurodegenerative diseases. In: Mediterranean Diet. Cham: Humana Press; 2016. pp. 153-164

[6] Plassman BL, Williams JW, Burke JR, Holsinger T, Benjamin S. Systematic review: Factors associated with risk for and possible prevention of cognitive decline in later life. Annals of Internal Medicine. 2010;**153**(3):182-193

[7] Sofi F. The Mediterranean diet revisited: Evidence of its effectiveness grows. Current Opinion in Cardiology. 2009;**24**(5) 442-446

[8] Esposito E, Rotilio D, Di Matteo V, Di Giulio C, Cacchio M, Algeri S. A review of specific dietary antioxidants and the effects on biochemical mechanisms related to neurodegenerative processes. Neurobiology of Aging. 2002;**23**(5):719-735

[9] Trichopoulou A, Costacou T, Bamia C, Trichopoulos D. Adherence to a Mediterranean diet and survival in a Greek population. The New England Journal of Medicine. 2003;**2003**(348):2599-2608

[10] Panagiotakos DB, Pitsavos C, Arvaniti F, Stefanadis C. Adherence to the Mediterranean food pattern predicts the prevalence of hypertension, hypercholesterolemia, diabetes and obesity, among healthy adults; the accuracy of the MedDietScore. Preventive Medicine. 2007;**44**(4):335-340

[11] Valls-Pedret C, Sala-Vila A, Serra-Mir M, Corella D, de la Torre R, Martínez-González MÁ. et al. Mediterranean diet and age-related cognitive decline: A randomized clinical trial. JAMA Internal Medicine. 2015;**175**(7):1094-1103

[12] Ngandu T, Lehtisalo J, Solomon A, Levälahti E, Ahtiluoto S, Antikainen R, et al. A 2 year multidomain intervention of diet, exercise, cognitive training, and vascular risk monitoring versus control to prevent cognitive decline in at-risk elderly people (FINGER): A randomised controlled trial. The Lancet. 2015;**385**(9984):2255-2263

[13] Tosti V, Bertozzi B, Fontana L. Health benefits of the Mediterranean diet: Metabolic and molecular mechanisms. The Journals of Gerontology: Series A. 2017;glx227

[14] Solfrizzi V, Custodero C, Lozupone M, Imbimbo BP, Valiani V, Agosti P, et al. Relationships of dietary patterns, foods, and micro-and macronutrients with Alzheimer's disease and late-life cognitive disorders: A systematic review. Journal of Alzheimer's Disease. 2017;**59**(3):815-849

[15] Olivera-Pueyo J, Pelegrín-Valero C. Dietary supplements for cognitive impairment. Actas Españolas de Psiquiatría. 2017;**45**(Supplement):37-47

[16] Miranda A, Gomez-Gaete C, Mennickent S. Role of Mediterranean diet on the prevention of Alzheimer disease. Revista Médica de Chile. 2017;**145**(4):501-507

[17] Anastasiou CA, Yannakoulia M, Kosmidis MH, Dardiotis E, Hadjigeorgiou GM, Sakka P, et al. Mediterranean diet and cognitive health: Initial results from the hellenic longitudinal investigation of ageing and diet. PLoS One. 2017;**12**(8):e0182048

[18] Hardman RJ, Meyer D, Kennedy G, Macpherson H, Scholey AB, Pipingas A. The association between adherence to a Mediterranean style diet and cognition in older people: The impact of medication. Clinical Nutrition. 2017 (in press)

[19] Caruana M, Cauchi R, Vassallo N. Putative role of red wine polyphenols against brain pathology in Alzheimer's and Parkinson's disease. Frontiers in Nutrition. 2016;**3**:31

[20] Takaoka M. Of the phenolic substances of white hellebore (*Veratrum grandiflorum* Loes. fil.). Journal of the Faculty of Science, Hokkaido University. 1940;**3**:1-16

[21] Langcake P, Pryce RJ. The production of resveratrol by *Vitis vinifera* and other members of the Vitaceae as a response to infection or injury. Physiological Plant Pathology. 1976;**9**(1):77-86

[22] Walle T. Bioavailability of resveratrol. Annals of the New York Academy of Sciences. 2011;**1215**(1):9-15

[23] Ahmed T, Javed S, Javed S, Tariq A, Šamec D, Tejada S, et al. Resveratrol and Alzheimer's disease: Mechanistic insights. Molecular Neurobiology. 2017;**54**(4):2622-2635

[24] Fernández-Mar MI, Mateos R, García-Parrilla MC, Puertas B, Cantos-Villar E. Bioactive compounds in wine: Resveratrol, hydroxytyrosol and melatonin: A review. Food Chemistry. 2012;**130**(4):797-813

[25] Shi J, He M, Cao J, Wang H, Ding J, Jiao Y, Li R, He J, et al. The comparative analysis of the potential relationship between resveratrol and stilbene synthase gene family in the development stages of grapes (*Vitis quinquangularis* and *Vitis vinifera*). Plant Physiology and Biochemistry. 2014;**74**:24-32

[26] Guerrero RF, Garcia-Parrilla MC, Puertas B, Cantos-Villar E. Wine, resveratrol and health: A review. Natural Product Communications. 2009;**4**(5):635-658

[27] Marhuenda J, Medina S, Martínez-Hernández P, Arina S, Zafrilla P, Mulero J, et al. Melatonin and hydroxytyrosol-rich wines influence the generation of DNA oxidation catabolites linked to mutagenesis after the ingestion of three types of wine by healthy volunteers. Food & Function. 2016;**7**(12):4781-4796

[28] Marhuenda J, Medina S, Martínez-Hernández P, Arina S, Zafrilla P, Mulero J, et al. Melatonin and hydroxytyrosol protect against oxidative stress related to the central nervous system after the ingestion of three types of wine by healthy volunteers. Food & Function. 2017;**8**(1):64-74

[29] Marhuenda J, Medina S, Martínez-Hernández P, Arina S, Zafrilla P, Mulero J, et al. Effect of the dietary intake of melatonin-and hydroxytyrosol-rich wines by healthy female volunteers on the systemic lipidomic-related oxylipins. Food & Function. 2017;**8**(10):3745-3757

[30] Lyons MM, Yu C, Toma R, Cho SY, Reiboldt W, Lee J, van Breemen RB. Resveratrol in raw and baked blueberries and bilberries. Journal of Agricultural and Food Chemistry. 2003;**51**(20):5867-5870

[31] Rimando AM, Kalt W, Magee JB, Dewey J, Ballington JR. Resveratrol, pterostilbene, and piceatannol in vaccinium berries. Journal of Agricultural and Food Chemistry. 2004;**52**(15):4713-4719

[32] Huang X, Mazza G. Simultaneous analysis of serotonin, melatonin, piceid and resveratrol in fruits using liquid chromatography tandem mass spectrometry. Journal of Chromatography. A. 2011;**1218**(25):3890-3899

[33] Counet C, Callemien D, Collin S. Chocolate and cocoa: New sources of trans-resveratrol and trans-piceid. Food Chemistry. 2006;**98**(4):649-657

[34] Chiva-Blanch G, Urpi-Sarda M, Rotchés-Ribalta M, Zamora-Ros R, Llorach R, Lamuela-Raventós RM, Estruch R, Andrés-Lacueva C. Determination of resveratrol and piceid in beer matrices by solid-phase extraction and liquid chromatography– tandem mass spectrometry. Journal of Chromatography. A. 2011;**218**(5):698-705

[35] Ares AM, González Y, Nozal MJ, Bernal JL, Higes M, Bernal J. Development and validation of a liquid chromatography with mass spectrometry method to determine resveratrol and piceid isomers in beeswax. Journal of Separation Science. 2015;**38**(2):197-204

[36] Sales JM, Resurreccion AV. Resveratrol in peanuts. Critical Reviews in Food Science and Nutrition. 2014;**54**(6):734-770

[37] Lu Q, Zhao Q, Yu QW, Feng Y. Use of pollen solid-phase extraction for the determination of trans-resveratrol in peanut oils. Journal of Agricultural and Food Chemistry. 2015;**63**(19):4771-4776

[38] de Oliveira MR, Chenet AL, Duarte AR, Scaini G, Quevedo J. Molecular mechanisms underlying the anti-depressant effects of resveratrol: A review. Molecular Neurobiology; 2017. pp. 1-17

[39] Sinclair D. Sirtuins for healthy neurons. Nature Genetics. 2005;**37**(4):339-340

[40] Porquet D, Casadesus G, Bayod S, Vicente A, Canudas AM, Vilaplana J, Pelegrí C, Sanfeliu C, et al. Dietary resveratrol prevents Alzheimer's markers and increases life span in SAMP8. Age. 2013;**35**(5):1851-1865

[41] Porquet D, Grinan-Ferre C, Ferrer I, Camins A, Sanfeliu C, Del Valle J, Pallàs M. Neuroprotective role of trans-resveratrol in a murine model of familial Alzheimer's disease. Journal of Alzheimer's Disease. 2014;**42**(4):1209-1220

[42] Parker JA, Arango M, Abderrahmane S, Lambert E, Tourette C, Catoire H, Néri C. Resveratrol rescues mutant polyglutamine cytotoxicity in nematode and mammalian neurons. Nature Genetics. 2005;**37**(4):349-350

[43] Karlsson J, Emgard M, Brundin P, Burkitt MJ. Trans-resveratrol protects embryonic mesencephalic cells from tert-butyl hydroperoxide. Journal of Neurochemistry. 2000;**75**(1): 141-150

[44] Karuppagounder SS, Pinto JT, Xu H, Chen HL, Beal MF, Gibson GE. Dietary supplementation with resveratrol reduces plaque pathology in a transgenic model of Alzheimer's disease. Neurochemistry International. 2009;**54**(2):111-118

[45] Draczynska-Lusiak B, Chen YM, Sun AY. Oxidized lipoproteins activate NF-kappaB binding activity and apoptosis in PC12 cells. NeuroReport. 1998;**9**(3):527-532

[46] Sun AY, Draczynska-Lusiak B, Sun GY. Oxidized lipoproteins, beta amyloid peptides and Alzheimer's disease. Neurotoxicity Research. 2001;**3**(2):167-178

[47] Hardeland R, Pandi-Perumal SR. Melatonin, a potent agent in antioxidative defense: Actions as a natural food constituent, gastrointestinal factor, drug and prodrug. Nutrition and Metabolism. 2005;**2**(1):22

[48] Sprenger J, Hardeland R, Fuhrberg B, Han SZ. Melatonin and other 5-methoxylated indoles in yeast: Presence in high concentrations and dependence on tryptophan availability. Cytologia. 1999;**64**(2):209-213

[49] Waldhauser F, Waldhauser M, Lieberman HR, Deng MH, Lynch HJ, Wurtman RJ. Bioavailability of oral melatonin in humans. Neuroendocrinology. 1984;**39**(4):307-313

[50] Fourtillan JB, Brisson AM, Gobin P, Ingrand I, Decourt JP, Girault J. Bioavailability of melatonin in humans after day-time administration of D7 melatonin. Biopharmaceutics & Drug Disposition. 2000;**21**(1):15-22

[51] Di WL, Kadva A, Johnston A, Silman R. Variable bioavailability of oral melatonin. The New England Journal of Medicine. 1997;**336**(14):1028-1029

[52] EFSA (2010). Scientific opinion on the substantiation of health claims related to melatonin and subjective feelings of jet lag (ID 1953), and reduction of sleep onset latency, and improvement of sleep quality (ID 1953) pursuant to Article 13(1) of Regulation

[53] De la Puerta C, Carrascosa-Salmoral MP, García-Luna PP, Lardone PJ, Herrera JL, Fernandez-Montesinos R, et al. Melatonin is a phytochemical in olive oil. Food Chemistry. 2007;**104**(2):609-612

[54] García Parrilla MDC, Hornedo Ortega R, Cerezo López AB, Troncoso González AM, Mas A. Melatonin and other tryptophan metabolites produced by yeasts: Implications in cardiovascular and neurodegenerative diseases. Frontiers in Microbiology. 2016;**6**:1565

[55] Tan DX, Zanghi BM, Manchester LC, Reiter RJ. Melatonin identified in meats and other food stuffs: Potentially nutritional impact. Journal of Pineal Research. 2014;**57**(2):213-218

[56] Iriti M, Rossoni M, Faoro F. Melatonin content in grape: Myth or panacea? Journal of the Science of Food and Agriculture. 2006;**86**:1432-1438

[57] Mercolini L, Protti M, Saracino MA, Mandrone M, Antognoni F, Poli F. Analytical profiling of bioactive phenolic compounds in argan (*Argania spinosa*) leaves by combined microextraction by packed sorbent (MEPS) and LC-DAD-MS/MS. Phytochemical Analysis. 2016;**27**(1):41-49

[58] Fernández-Cruz E, Álvarez-Fernández MA, Valero E, Troncoso AM, García-Parrilla MC. Melatonin and derived L-tryptophan metabolites produced during alcoholic fermentation by different wine yeast strains. Food Chemistry. 2017;**217**:431-437

[59] Guenther AL, Schmidt SI, Laatsch H, Fotso S, Ness H, Ressmeyer AR, et al. Reactions of the melatonin metabolite AMK (N1-acetyl-5-methoxykynuramine) with reactive nitrogen species: Formation of novel compounds, 3-acetamidomethyl-6-methoxycinnolinone and 3-nitro-AMK. Journal of Pineal Research. 2005;**39**(3):251-260

[60] Schaefer M, Hardeland R. The melatonin metabolite N1-acetyl-5-methoxykynuramine is a potent singlet oxygen scavenger. Journal of Pineal Research. 2009;**46**(1):49-52

[61] Than NN, Heer C, Laatsch H, Hardeland R. Reactions of the melatonin metabolite N1-acetyl-5-methoxykynuramine (AMK) with the ABTS cation radical: Identification of new oxidation products. Redox Report: Communications in Free Radical Research. 2006;**11**(1):15-24

[62] Reiter RJ. Melatonin: Clinical relevance. Best Practice & Research Clinical Endocrinology & Metabolism. 2003;**17**(2):273-285

[63] Reiter RJ, Tan DX, Maldonado MD. Melatonin as an antioxidant: Physiology versus pharmacology. Journal of Pineal Research. 2005;**39**(2):215-216

[64] Herraiz T, Galisteo J. Endogenous and dietary indoles: A class of antioxidants and radical scavengers in the ABTS assay. Free Radical Research. 2004;**38**(3):323-331

[65] Nogués MR, Giralt M, Romeu M, Mulero M, Sánchez-Martos V, Rodríguez E, et al. Melatonin reduces oxidative stress in erythrocytes and plasma of senescence-accelerated mice. Journal of Pineal Research. 2006;**41**(2):142-149

[66] Scheer FAJL, Van Montfrans GA, Van Someren EJW, Mairuhu G, Buijs RM. Daily nighttime melatonin reduces blood pressure in male patients with essential hypertension. Hypertension. 2004;**43**(2I):192-197

[67] Karbownik M, Lewinski A, Reiter RJ. Anticarcinogenic actions of melatonin which involve antioxidative processes: Comparison with other antioxidants. International Journal of Biochemistry and Cell Biology. 2001;**33**(8):735-753

[68] Srinivasan V, Pandi-Perumal SR, Maestroni GJM, Esquifino AI, Hardeland R, Cardinali DP. Role of melatonin in neurodegenerative diseases. Neurotoxicity Research. 2005; 7(4):293-318

[69] Miller E, Morel A, Saso L, Saluk J. Melatonin redox activity. Its potential clinical applications in neurodegenerative disorders. Current Topics in Medicinal Chemistry. 2015;**15**(2):163-169

[70] Khaldy H, Escames G, Leon J, Vives F, Luna JD, Acuña-Castroviejo D. Comparative effects of melatonin, l-deprenyl, Trolox and ascorbate in the suppression of hydroxyl radical formation during dopamine autoxidation in vitro. Journal of Pineal Research. 2000;**29**(2):100-107

[71] van der Schyf CJ, Castagnoli K, Palmer S, Hazelwood L, Castagnoli N. Melatonin fails to protect against long-term MPTP-induced dopamine depletion in mouse striatum. Neurotoxicity Research. 1999;**1**(4):261-269

[72] Medeiros CAM, De Bruin PFC, Lopes LA, Magalhães MC, de Lourdes Seabra M, de Brui VMS. Effect of exogenous melatonin on sleep and motor dysfunction in Parkinson's disease. Journal of Neurology. 2007;**254**(4):459-464

[73] Kashani IR, Rajabi Z, Akbari M, Hassanzadeh G, Mohseni A, Eramsadati MK, et al. Protective effects of melatonin against mitochondrial injury in a mouse model of multiple sclerosis. Experimental Brain Research. 2014;**232**(9):2835-2846

[74] Miller E, Walczak A, Majsterek I, Kędziora J. Melatonin reduces oxidative stress in the erythrocytes of multiple sclerosis patients with secondary progressive clinical course. Journal of Neuroimmunology. 2013;**257**(1):97-101

[75] Zhou J, Zhang S, Zhao X, Wei T. Melatonin impairs NADPH oxidase assembly and decreases superoxide anion production in microglia exposed to amyloid-β1-42. Journal of Pineal Research. 2008;**45**(2):157-165

[76] Olcese JM, Cao C, Mori T, Mamcarz MB, Maxwell A, Runfeldt MJ, et al. Protection against cognitive deficits and markers of neurodegeneration by long-term oral administration of melatonin in a transgenic model of Alzheimer disease. Journal of Pineal Research. 2009;**47**(1):82-96

[77] He H, Dong W, Huang F. Anti-amyloidogenic and anti-apoptotic role of melatonin in Alzheimer disease. Current Neuropharmacology. 2010;**8**(3):211-217

[78] Das A, Wallace G, Reiter RJ, Varma AK, Ray SK, Banik NL. Overexpression of melatonin membrane receptors increases calcium-binding proteins and protects VSC4. 1 motoneurons from glutamate toxicity through multiple mechanisms. Journal of Pineal Research. 2013;**54**(1):58-68

[79] Weishaupt JH, Bartels C, Pölking E, Dietrich J, Rohde G, Poeggeler B, et al. Reduced oxidative damage in ALS by high-dose enteral melatonin treatment. Journal of Pineal Research. 2006;**41**(4):313-323

[80] Zhang Y, Cook A, Kim J, Baranov SV, Jiang J, Smith K, et al. Melatonin inhibits the caspase-1/cytochrome c/caspase-3 cell death pathway, inhibits MT1 receptor loss and delays disease progression in a mouse model of amyotrophic lateral sclerosis. Neurobiology of Disease. 2013;**55**:26-35

[81] Liu G, Aliaga L, Cai H. α-synuclein, LRRK2 and their interplay in Parkinson's disease. Future Neurology. 2012;7(2):145-153

[82] Mateos R, Espartero JL, Trujillo M, Rios JJ, León-Camacho M, Alcudia F, Cert A. Determination of phenols, flavones, and lignans in virgin olive oils by solid-phase extraction and high-performance liquid chromatography with diode array ultraviolet detection. Journal of Agricultural and Food Chemistry. 2001;**49**(5):2185-2192

[83] Di Tommaso D, Calabrese R, Rotilio D. Identification and quantitation of hydroxytyrosol in Italian wines. Journal of Separation Science. 1998;**21**(10):549-553

[84] Boselli E, Minardi M, Giomo A, Frega NG. Phenolic composition and quality of white doc wines from Marche (Italy). Analytica Chimica Acta. 2006;**563**(1):93-100

[85] Dudley JI, Lekli I, Mukherjee S, Das M, Bertelli AA, Das DK. Retraction: Does white wine qualify for french paradox? comparison of the cardioprotective effects of red and white wines and their constituents: Resveratrol, tyrosol, and hydroxytyrosol. Journal of Agricultural and Food Chemistry. 2012;**60**(10):2767-2767

[86] Proestos C, Bakogiannis A, Psarianos C, Koutinas AA, Kanellaki M, Komaitis M. High performance liquid chromatography analysis of phenolic substances in Greek wines. Food Control. 2005;**16**(4):319-323

[87] Minuti L, Pellegrino RM, Tesei I. Simple extraction method and gas chromatography–mass spectrometry in the selective ion monitoring mode for the determination of phenols in wine. Journal of Chromatography A. 2006;**1114**(2):263-268

[88] de la Torre R, Covas MI, Pujadas MA, Fitó M, Farré, M. Is dopamine behind the health benefits of red wine?. European Journal of Nutrition. 2006;**45**(5):307-310

[89] Schröder H, de la Torre R, Estruch R, Corella D, Martínez-González MA, Salas-Salvadó J, et al. Alcohol consumption is associated with high concentrations of urinary hydroxytyrosol. The American Journal of Clinical Nutrition. 2009;**90**(5):1329-1335

[90] Cornwell DG, Ma J. Nutritional benefit of olive oil: The biological effects of hydroxytyrosol and its arylating quinone adducts. Journal of Agricultural and Food Chemistry. 2008;**56**(19):8774-8786

[91] Goya L, Mateos R, Bravo L. Effect of the olive oil phenol hydroxytyrosol on human hepatoma HepG2 cells. European Journal of Nutrition. 2007;**46**(2):70-78

[92] Bazoti FN, Bergquist J, Markides KE, Tsarbopoulos A. Noncovalent interaction between amyloid-β-peptide (1-40) and oleuropein studied by electrospray ionization mass spectrometry. Journal of the American Society for Mass Spectrometry. 2006;**17**(4):568-575

[93] González-Correa JA, Navas MD, Lopez-Villodres JA, Trujillo M, Espartero JL, De La Cruz JP. Neuroprotective effect of hydroxytyrosol and hydroxytyrosol acetate in rat brain slices subjected to hypoxia–reoxygenation. Neuroscience Letters. 2008;**446**(2):143-146

[94] Pandey KB, Rizvi SI. Plant polyphenols as dietary antioxidants in human health and disease. Oxidative Medicine and Cellular Longevity. 2009;**2**(5):270-278

[95] Du Y, Guo H, Lou H. Grape seed polyphenols protect cardiac cells from apoptosis via induction of endogenous antioxidant enzymes. Journal of Agricultural and Food Chemistry. 2007;**55**(5):1695-1701

[96] Milner JA. Reducing the risk of cancer. In: Functional Foods. Springer US; 1994. pp. 39-70

[97] Lee HH, Itokawa H, Kozuka M. Asian herbal products: The basis for development of high-quality dietary supplements and new medicines. In: Shi J, Ho CT, Shahidi F, editors. Asian Functional Foods. 2005:21-72

[98] Karbowniczek A, Wierzba-Bobrowicz T, Mendel T, Nauman P. Cerebral amyloid angiopathy manifested as a brain tumour. Clinical and neuropathological characteristics of two cases. Folia Neuropathologica. 2012;**50**(2):194-200

[99] Geiser RJ, Chastain SE, Moss MA. Regulation of Alzheimer's disease-associated mRNA expression by green tea catechins and black tea theaflavins. Alzheimer's & Dementia: The Journal of the Alzheimer's Association. 2017;**13**(7):274-275

[100] Spencer JP. The impact of fruit flavonoids on memory and cognition. British Journal of Nutrition. 2010;**104**(S3):S40-S47

[101] Guasch-Ferré M, Merino J, Sun Q, Fitó M, Salas-Salvadó J. Dietary polyphenols, Mediterranean diet, prediabetes, and Type 2 diabetes: A narrative review of the evidence. Oxidative Medicine and Cellular Longevity. 2017;**2017**

[102] Cho J, Kang JS, Long PH, Jing J, Back Y, Chung KS. Antioxidant and memory enhancing effects of purple sweet potato anthocyanin and cordyceps mushroom extract. Archives of Pharmacal Research. 2003;**26**(10):821-825

[103] Strathearn KE, Yousef GG, Grace MH, Roy SL, Tambe MA, Ferruzzi MG, et al. Neuroprotective effects of anthocyanin-and proanthocyanidin-rich extracts in cellular models of Parkinson' s disease. Brain Research. 2014;**1555**:60-77

[104] Kelsey N, Hulick W, Winter A, Ross E, Linseman D. Neuroprotective effects of anthocyanins on apoptosis induced by mitochondrial oxidative stress. Nutritional Neuroscience. 2011;**14**(6):249-259

Chapter 3

Development and Characterization of Fish-Based Superfoods

Abstract

The importance of superfoods has been well recognized in connection with health promotion, disease risk reduction, and reduction in health care costs. Fish processing generates large quantities of the by-products which are usually discarded. However, these by-products contain nutritious protein and ω-3 rich oils. Isoelectric solubilization/precipitation (ISP) is a relatively new method that can be used to recover fish protein isolate (FPI) from fish processing by-products or other low-value meat materials. FPI can be used as a main ingredient in the development of superfoods with functional ingredients such as ω-3 rich oils, dietary fiber, and salt substitute. These functional ingredients have demonstrated health benefits especially for cardiovascular disease. Therefore, this book chapter focuses on the development of superfoods from ISP-recovered FPI by incorporating such ingredients as ω-3 oil, dietary fiber, and salt substitute.

Keywords: superfoods, fish, isoelectric solubilization/precipitation, ω-3 oil, dietary fiber, salt substitute

1. Introduction

Consumers' increasingly awareness and demand for healthful foods has hit a global status, thus, requiring appropriate and timely response from stakeholders in the food processing field. Beyond the nutritional values from food intake, consumers are looking for food products that would doubly provide them with health benefits. Among other factors, higher health care cost, recent developments in scientific discoveries linking dietary habits with many diseases can be attributed to this increasing demand for functional foods by consumers [1].

Market trend in functional foods and superfoods is soaring and projections show exponential increase in demands in the future.

Colloquially, superfoods have been described as foods that are appealing and are able to deliver more calories per bite [2]. Superfoods have indigenously been referred to as functional foods [3], even now, there is still a thin line separating the two definitions. Several phrases have been used to define functional foods and these definitions vary accordingly, from continent to continent. For instance, in Europe, "a food product can only be considered functional if together with the basic nutritional impact it has beneficial effects on one or more functions of the human organism thus either improving the general and physical conditions or/and decreasing the risk of the evolution of diseases" [4]. Conversely, in the US, functional food is defined as "food or food ingredient that may provide a health benefit beyond the traditional nutrients it contains [5]. According to Siró et al. [4], this is not a legal definition and the US has not defined functional foods for regulatory purposes. Tahergorabi et al. [6] gave a comprehensive description of what functional food should possess. They characterized functional foods as modified compositions and/or processing conditions to prevent or limit the presence of certain potentially harmful components to human health, and/or inclusion of certain desirable substances, present either naturally or added, with proven health benefits. The fundamental implication from this pool of definitions is that, functional foods have characteristic properties, which in addition to their nutritive qualities, provide health benefits when consumed. The many purported health benefits associated with superfoods/functional foods can be attributed to compounds found in these foods [1]. Bioactive compounds found in both plants and animals are responsible for the many health claims of superfoods and functional foods.

In commercial fish processing, about 60–70% of the live fish weight are probably discarded as processing by-products with just a little percentage being marketed as fillet [7]. Fish processing by-products are the remainder of the fish after processing that are not considered as fitting for human consumption, and it includes the frames, heads, skins, bones, scales, visceral from the fish. They are sometimes used in animal feeds and fertilizers, or ultimately at land filled sites. These "neglected" fish parts present opportunity to recover useful macromolecules that would have otherwise been discarded. Fish processing by-products are good sources of protein and Fat. They contain bioactive compounds such as proteins, peptides, amino acids [8] and fatty acids. Peptides are produced from hydrolysis of protein. Peptides are known to have antioxidative, antihypertensive and antithrombotic effects [9]. Fish oils produced from marine fish such as salmon, cod, and sardine which are the common sources of omega-3 polyunsaturated fatty acid (ω-3 PUFA). It is known that, ω-3 PUFA oil and the proteins are the bioactive compound that gives fish its "superfood" status [6]. The main ω-3 PUFAs found in aquatic animals are eicosapentaenoic acid (EPA, 20:5 ω3), and docosahexaenoic acid (DHA, 22:6 ω3), while linoleic acid (La, 18:2 ω6) and arachidonic acid (AA, 20:4 ω6) are the main ω-6 PUFA in aquatic animals [10]. US Food and Drug Administration (FDA) as well as, the European Food Safety Authority approved a health claim for reduced risk of cardiovascular diseases (CVDs), and if so, sudden death [11] for foods containing omeg-3 PUFAS, especially EPA and DHA. EPA and DHA are known to have antioxidant properties [12].

The benefits derived from superfoods/functional foods can be obtained not just from natural foods but from novel foods as well [1]. This chapter will succinctly review the development of functional foods from seafood protein and other functional ingredients.

2. Recovery of proteins from fish processing by-products

There are several methods for recovering protein from fish. The method of recovery affects the quality of the protein and oil, and their functionality. Fish protein isolates (FPI), fish protein hydrolysates (FPH), and surimi among others are resulting products of the various methods used in the recovery process. FPH can be produced by breaking down of the peptide bonds of the fish protein using proteolytic enzymes [13], as well as chemical methods. FPH has been used in developing many products with functional properties. For instance, the resulting protein concentrate can be dried to form a stabilized product, fish protein powder (FPP) [14]. Formulations and ready-to eat foods have been developed from FPP. However, the cost of production of FPP has placed a limit on its use [14]. Additionally, one grievous concerned about foods developed from FPH is off flavor and bitter taste associated with these foods [10, 12].

Surimi (**Figure 1**) is a concentrate of myofibrillar protein from minced fish flesh that has been deboned mechanically, and washed. It has been a long principal food ingredient in Japan diets, and has now become an integral part of many other countries diet [15, 16]. However, surimi manufacturing cannot be applied to fish processing by-products. If used, the resultant protein will have poor texture, off-odor and off-color [17]. Novel extractive technology which uses pH at basic and acidic ranges to solubilize and precipitate the protein is currently been used and it is patented [18]. Protein recovery using pH-shift also known as isoelectric solubilization/precipitation (ISP) has recently been used in developing fish protein products. ISP processing allows selective and efficient recovery of nutritious protein muscles, separation of lipids and removal of materials that are not intended for human consumption such as bone, scales, skin, etc. [19]. FPIs

Figure 1. A block of frozen surimi (usually formed in 10 kg blocks).

have been recovered using ISP both in a batch mode at the laboratory scale [20, 21] and on pilot scale [22]. ISP has also recently been used to recover protein muscle from chicken meat [6]. FPI is a concentrated fish protein that contains 90% of the dry material as protein content. FPI has constantly being used in literature to mean muscle protein that has been recovered with protein shift (ISP). It is uncooked and definitely not eaten in its form but like FPP can be used in formulation of many food products. The overall process of fish muscle recovery using ISP involves solubilization and precipitation using a pH known as Isoelectric point (pI).

The pI is specific and differs from protein to protein, and isoelectric focusing is often done to identify the pI. At this pH, the protein ions exist in equilibrium, that is they have zero net charge and are called zwitterions. The protein side chain can assume different electrostatic charges depending on the condition it is subjected to. This comes to suggest that, change in environment of the fish protein affects its solubility. That is, fish muscle solubility can be "turned" on or "turned" off depending on the surrounding environment that they are placed in. Addition of an acid to a solution leads to the dissociation of the acid producing hydronium (H_3O^+). At this low pH, the negatively charged side chains on glutamyl or aspartyl receive the hydronium and becomes protonated. This makes the solution more positively charged. When a base is added, it as well dissociates giving off hydroxide ion (^-OH). The side chains on tyrosyl, tryptophanyl, cysteinyl, lysyl, argininyl or histidinyl residues become deprotonated by giving off its hydrogen ion. This makes the surface of the solution more negatively charged. This has been attributed to the solubilization of fish muscle protein during protonation of glutamyl and aspartyl (pKa = 3.8 and 4.2 respectively) residue at acidic pH and the subsequent deprotonation of tyrosyl, tryptophanyl, and cysteinyl, (pKa = 9.5–10.5, 9.1–10.8, and 9.1–10.8, respectively) residue at basic pH. When equilibrium is reached, and protein solution attains homeostasis, the final status of a protein surface electrostatic charge at a given pH is referred to as the net charge. The accumulation of net positive or negative charge encourages protein–protein electrostatic repulsion and increased hydrodynamic volume due to expansion and swelling [23]. Protein–protein hydrophobic interaction decreases as the protein-water interaction increases. As a result of protein becoming more polar, the protein surface become surrounded with water, making it water soluble. It is however, likely to adjust the pH of a protein solution so that the number of negative charges and positive charges balances, and hence, the protein molecule assumes a net zero electrostatic charge.

The above description forms the theoretical substance for isoelectric solubilization/precipitation (ISP) that allows mechanistic understanding of pH-induced protein solubility and precipitation. The ability to recover a protein from any animal sources including fish is generally dependent on the knowledge of its pI. The initial ISP recovery process, encompasses homogenization of the fish to release the protein from the muscles and subsequent solubilization and precipitation using a pH known as Isoelectric point (PI). The homogenate is transferred into a beaker and the pH is adjusted to 11.50 ± 0.05 with 10 and 1 N NaOH. Solubilization of the protein occurs at a basic pH of 11.50 and it is then followed by centrifugation, resulting in three layers; top deposit of fish oil, middle layer of fish muscle protein solution, and bottom deposit of insoluble (bones, skin, scale, insoluble protein, membrane lipids, etc.). The middle fish protein layer is collected and pH adjusted to 5.5 ± 0.05 with 10 and 1 N HCl to allow the protein to precipitate. It is centrifuged into two layers: top, process water; bottom layer, precipitate-fish protein isolate. The summarized process is presented in **Figure 2**.

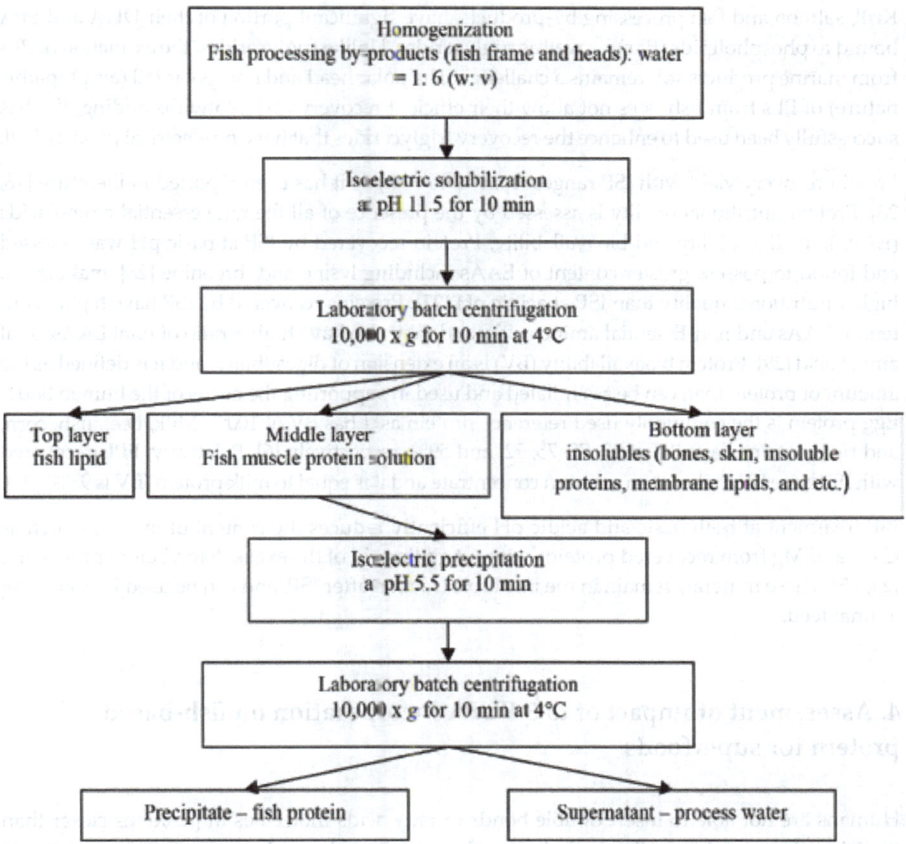

Figure 2. Flow diagram for fish protein isolate recovery using isoelectric solubilization/precipitation. Reproduced from Tahergorabi et al. [40].

3. Nutritional properties of recovered proteins by isoelectric solubilization/precipitation

Isoelectric solubilization/precipitation allows efficient and selective recovery of fish proteins and oil without changing the nutritional and functional value of the food products [19]. Evidentially, this process of recovering protein and fats does not affect the composition of ω-3 or ω-6 PUFA, and consequently, the ω-3/ω-6 ratio [24]. Also, the fatty acid composition of the fish is not altered after it has been recovered with ISP at both basic and acidic pH, as it remains fairly the same as in the starting fish [25]. The other known processes and conditions, unlike ISP, involves heating which affect protein degradation as well as endogenous antioxidants and fatty acids [26]. Hence, protein recovered by ISP significantly reduces protein degradation and fat oxidation.

Krill, salmon and fish processing by-products, have significant portion of their DHA and EPA bound to phospholipids (PLs), as well as triglycerides. Unlike triglycerides, the extraction of PLs from marine products still remains a challenge. The polar head and non-polar tail (amphipathic nature) of PLs from fish does not allow their efficient recovery [27]. Notwithstanding, ISP has successfully been used to enhance the recovery triglycerides that have no electrical polarity [19].

Protein recovery yield with ISP ranges from 42 to 90%, as it has been reported in literature [18, 23]. Protein nutritional quality is assessed by the presence of all the nine essential amino acids (EAAs), its digestibility and bioavailability. Protein recovered by ISP at basic pH was assessed and found to possess greater content of EAAs including lysine and threonine [28], making it a higher nutritional quality than ISP at acidic pH [27]. Proteins recovered by ISP have higher content of EAAs and non-Essential amino acids, and they also have higher ratio of total EAAs: total amino acid [29]. Protein bioavailability (BV) is an extension of digestibility, and it is defined as the amount of protein than can be assimilated and used in supporting the needs of the human body. Egg protein is the commonly used reference protein as it has BV of 100%. Milk, beef, fish, corn, and rice proteins have BV of 93, 75, 75, 72, and 59%, respectively [6]. Relatively, FPIs recovered with ISP have BV higher than soybean concentrate and it is equal to milk protein (BV is 93%) [30].

ISP treatment at both basic and acidic pH efficiently reduces the content of minerals such as Ca, P and Mg from recovered proteins without getting rid of the exoskeleton before processing [24, 25]. These minerals remain in the insoluble fraction after ISP, and can be used in preparing animal feed.

4. Assessment of impact of ω-3 PUFA incorporation on fish-based protein for superfoods

Humans are not able to insert double bonds in fatty acids molecules in positions closer than the 7th carbon bond from the methyl group because they do not have the necessary enzymes [31, 32]. Fish can be classified based on whether it stores its lipid in the liver or in the flesh. The former and latter groups of fish are known as lean fish (e.g. Cod fish) and fatty fish (e.g., Mackerel, tuna, salmon), respectively [33]. The amount of oil varies among fish in the same class. Fatty fish have comparatively higher amount of DHA and EPA oils than lean fish. ω-3 PUFAs are made up of long chains of carbon atoms with a methyl group at one and an acid group at the other end. They can be classified as saturated, as with no double carbon–carbon bonds, or unsaturated, that is with at least one double carbon–carbon bond. The position of the first double carbon–carbon bond from the methyl group end of an unsaturated fatty acid determines whether the saturated fatty acid would be called ω-3 or ω-6. ω-6 fatty acids, thus, the first double carbon–carbon bond is six carbon atoms away from the methyl group of the fatty acid chain. Similarly, ω-3 PUFA has the first double carbon–carbon bond three carbon atoms away from the methyl end. ω-3 PUFAS contains two healthy oils namely; EPA and DHA [34]. The third form ω-3 PUFA, ALA (α-linolenic acid) is predominantly obtained from plants sources such as leafy vegetables, flaxseeds, walnuts. ALA can be converted to long chain EPA and DHA [35] but synthesis of EPA and DHA from ALA is characterized by low amounts. Gender factor also affects conversion capacity of ALA to EPA and DHA. Women are able to synthesize more EPA and DHA from ALA because of estrogen effects as compared to men [10]. EPA and DHA can help in the protection against cardiovascular diseases (CVD) [36].

The western diet is typified by high amount of ω-6 PUFA, and trans-fatty acids dietary intake as well as decreased intake of ω-3 PUFA. A disproportionate ratio of ω-6 to ω-3 is at 15–20 to 1 as against the recommended 1 to 1 [37–39]. The current trend is to take dietary ω-3 PUFA supplement in the form of pill or capsule to make up for the large gap. Fish is not only known to provide a good source of protein (18–25% protein) but most varieties of fish are also low in cholesterol (15–25% mg/100 g), making fish highly suitable source of ω-3 PUFAs [12]. There is therefore, a logical justification for the fortification of fish-based protein to increase the content of dietary ω-3 fatty acid.

4.1. ω-3 PUFAs content in fish-based superfoods fortified with ω-3 PUFAs rich oil

Fat contents of FPI recovered with ISP are woefully low, usually below 2/100 g ("as-is" basis), and with this, meeting the recommended dietary intake of ω-3 PUFAs will be difficult. Tahergorabi et al. [40] examined the impact of fortification of FPI gels recovered with ISP using ω-3 PUFAs-rich oils. The oils were obtained from both plant and animals sources including flaxseed, fish, algae, krill, or their blend (Flaxseed, algae: fish, 8:1:1). These oils are good sources of ω-3 PUFAs, and hence their selection [41, 42]. The pastes were formulated with 10/100 g of ω-3 PUFAs- rich oils before cooking to form gels. The fortification increased ($P > 0.05$) the total content of ω-3 PUFAs (34–51%) of the gels, higher than gels that were not fortified (20%). The highest ($P > 0.05$) ω-3 PUFAs content was in gels fortified with flaxseed (51%), followed by blend (49%), krill (46%), and fish oil (34%). The EPA, DHA, ALA contents increment was characterized by variations in the fatty acid (FA) in total fatty acids. Krill and fish oil- fortified gels resulted in the greatest ($P > 0.05$) content of EPA (24 and 16%, respectively), with other fortified gels containing less than 3% of EPA. Results obtained from krill and fish oil are promising because, both oils are very good sources of EPA and DHA whereas algal oils contain DHA as their main ω-3 PUFA [43].

The differences in FA compositions of the gels resultantly affect the ratios of ω-6/ω-3 and unsaturated/saturated FAs (UFAs/SFAs). The lowest ω-6/ω-3 FAs ratio was seen in gels fortified with algae, krill, and fish oils (0.07, 0.11, and 0.12, respectively), followed by gels with added blend and flaxseed oil (0.29 and 0.32, respectively). The highest ($P < 0.05$) ratio of ω-6/ω-3 FA was seen in gels fortified with flaxseed and blend due to the high content of LA (linolenic acid) in flaxseed as compared to much lower LA content in the other oils. Despite the recorded highest ratio of ω-6/ω-3 with flaxseed oil, it is still much lower than the recommended 1/1 ratio [44]. It is still yet better to consume fish protein isolates fortified with any of the five oil types used in the present study, more advisably those fortified with flaxseed oil. This would help lessen the gap between ω-6 and ω-3 PUFAs in Western diets. On the other hand, gels fortified with flaxseed and blend showed the highest ($P < 0.05$) UFAs/SFAs ratio, 8.0 and 6.1 respectively; and the lowest ($P < 0.05$) ratio was recorded for fortification with krill, algae, and fish oils, 3.8, 3.5, and 2.3, respectively. Although the latter three oils resulted in lower UFAs/SFAs ratios in the gels, they still contained 2–4 times as much UFAs as SFAs.

4.2. Color properties of fish-based superfoods fortified with ω-3 PUFAs rich oils

The assessment of physicochemical properties of a new developed protein product is necessary for consumer acceptability. Consumers are tuned to certain original physical and chemical properties of proteins and are likely to reject products that do not meet their requirements. In the same vain, consumers may reject a product irrespective of its quality, if it does not appeal to their senses. Maintaining high sensory attributes has been one of the challenges confronting the development of functional and superfoods. Color as an attribute of fish and fish products is a strong determinant of consumers' acceptability. In as much as FPI recovered with ISP possess quality characteristic properties, its esthetic appearance has been a rough edge that needs to sharpened.

Whiteness of color in sea-foods is a desirable attribute and variable when it comes to quality assessment of such foods. In spite of the many touted qualities of protein and oil recovered by ISP, heat set gels made from fish protein isolate develop poor color due to dark pigments that are extracted and recovered with the proteins [45]. These dark pigment results in high yellowness (b*), thus, lowering the whiteness of the heat set gel. Numerous research studies have been geared towards improving the whitening attribute of heat set gels developed from proteins. Titanium dioxide (TiO_2) is a well-known whitening agent used in the cosmetic and food industry. Titanium dioxide has been used to improve the color attribute of gels developed from proteins recovered by ISP, in fish [46] and chicken proteins [47]. It does so by blocking/scattering light and giving white appearance [48]. Gels that have been treated with TiO_2 have color attribute improved beyond that of products that are not treated. Incorporation of ω-3 fatty acid into heat set gels developed from FPI have yielded positive results, improving color attributes. This discovery simply implies that, fortifying FPI with ω-3 PUFAs contributes to color improvement without having to add TiO_2. Fish protein isolate pastes were fortified with different types of ω-3 PUFAs oils, and gels subsequently were prepared [49]. The quest was to improve the color, whilst concurrently improving their nutrition and texture. The same observation was made by Pérez-Mateos et al. [50] when they fortified surimi with ω-3 from three different sources. Usually, vegetable oils are used in surimi-based products for color improvement [51]. Oils addition to products proportionally increase their lightness (L*) because of light scattering, thus, improving whiteness [50]. In measuring the whiteness of gel, values for the CIE (commission international d'Eclairage of France) color system using L*, a* (redness), and b* tristimulus color are determined and used to calculate the whiteness of the gel. Gel whiteness is calculated by the following equation [52]. The color of the gels was generally improved after it was formulated with ω-3 PUFAs oils.

4.3. Texture properties of fish-based superfoods fortified with ω-3 PUFAs rich oils

The texture of fish and fish product is a salient physicochemical property that determines the quality of fish and fish products. Texture measures the mechanical properties in the form hardness/firmness, resilience, cohesiveness, springiness, adhesiveness, and viscosity by vision, hearing, somesthesis, and kinesthesis of human sense [53]. Torsion test, Kramer shear test, and texture profile analysis (TPA) test are three possible methods that can be used to determine texture. Torsion test is the basic test due to its objectivity in measuring mechanical properties of protein-based gels [54]. Critical quality parameters for restructured gelled products are gel strength and cohesiveness. Cohesiveness as defined as "the extent to which a material can be deformed before it ruptures" and hardness as the force necessary to attain a given deformation" [55]. Trout protein gel fortified with or without ω-3 PUFAs oils were compared. Heat-set gels were fortified with flaxseed, fish, and algae oil and texture of the gels were measured using texture profile analysis (TPA) [49]. There was a general improvement in gel texture, particularly with algae oil. These results indicate the possibility of developing superfood products from ISP recovered FPI with acceptable texture properties using ω-3 PUFAs oils.

5. Role of salt substitute and ω-3 PUFAs rich oil in fish-based superfoods

Dietary sodium chloride has been a major component of many food products. The major mineral, sodium, has been correlated with hypertension and many cardiovascular diseases. In spite of this revelation of adverse effect of consuming this sodium related salt on triggering CVDs, it still remains a major dietary component in diets because of its undeniable taste. Therefore, during food processing, the sodium can be reduced or replaced to reduce salt-related CVDs. Salt substitute (potassium) is currently being exploited as a potential substitute for sodium in diets as it provides a non-pharmacological approach in lowering blood pressure. Potassium, unlike sodium has antihypertensive properties and are much higher recommended maximum intake level than sodium (sodium—2300 mg/day, potassium—4700 mg/day) [56].

Potassium Chloride-based salt substitute was used to extract myofibrillar proteins to obtain fish protein paste, a procedure described by Jaczynski and Park [57]. The fish protein paste was then chopped at low speed for 5 min in a universal food processor. The level of the salt substitute was found to be optimal and similar to salt (NaCl) in terms of texture and a color development likewise the protein gelation and reduction of water activity in heat-set fish protein gels [49]. The salt substitute contained 68/100 g of KCl and L- lysine mono-hydrochloride and calcium stearate. The resulting concentration of fish protein isolate is equivalent to 20 g of NaCl per 1000 g batch. It was then fortified with different ω-3 PUFAs oils and the fish protein heat-set gels were subsequently prepared. Gels made with salt substitute with different ω-3 PUFAs oil were analyzed for their sodium and potassium content. Sodium content was reduced with concurrent increase in potassium content in the gel.

6. Physicochemical properties of fish-based superfoods fortified with dietary fiber

The European community defined dietary fiber as a carbohydrate polymer with three or more monomeric units, which are neither digested nor absorbed in the human intestine and according to the American Association of Cereal Chemists (AACC) defined it as an edible part of plants or analogous carbohydrates, that are resistant to digestion and absorption in the human small intestine and can be partially or completely fermented by bacteria. Oligosaccharides, polysaccharides, lignin, and other plant sources provide fiber. Dietary fiber intake protects individuals against CVDs, diabetes, hypertension, obesity, stroke and certain gastrointestinal diseases [58, 59]. The health benefits of dietary fiber have been made to be understood by many countries, and they have developed guidelines, and have also allowed for fortification of food products with dietary fiber. The fortification of food with dietary fiber has been imperative, and has been encouraged, because most diets in the western countries are deficient in dietary fiber. Daily dietary recommendation for most European countries and for countries like Australia, and the New Zealand and the USA are in the order 30–35 g for men and 25–32 g for women [60]. The Dietary Guidelines Advisory committee (2010), find out that the average intake among Americans is only 15 g. The same can be said about other countries, as having low average intake of dietary fiber. Additionally, dietary fiber possesses gelling characteristics that have technological implications in food manufacturing and final food product. Dietary fiber has wide range of functionalities such as gel-forming abilities, cryoprotectant, thickener and stabilizer [61]. Cellulose is a polymer of linear glucose monomer (β 1–4 linkage) chain and it is the commonly used dietary fiber in food fortification. It is added in foods in powdered form. Powered cellulose is used in food fortification, and it has been used in baked goods as non-caloric bulk agent. Long chain-chain cellulose (> 110 μm), due to their porous nature, retains more water and oil than short chain fibers.

Naturally, the addition of dietary fiber to fish protein (surimi) is not common, and thus, limited literature report on it. Notwithstanding, soluble fibers such as carrageen, chicory root insulin, garrofin, guar, and xanthan have previously been added to surimi [62, 63]. The use of these soluble fibers resulted in loss of gel elasticity and strength coupled with gel hardening, and increase in brittleness of surimi protein [64]. This is not the case of surimi fortified with powdered cellulose as they tend to improve thermal gelation of surimi proteins [65]. In the fortification of surimi with powdered cellulose powder, the protein concentration and moisture content were maintained constant but variable concentrations of insoluble fiber, and silicon dioxide (SiO_2) were added to the surimi paste as an inert filler [66]. Insoluble fiber and SiO_2 were then added to the surimi pastes to a final total concentration of 8/100 g. The ingredients were chopped to ensure thorough mixing. The resulting pastes were subjected to texture analysis, color analysis, etc. Textural properties (shear stress and strain) of surimi fortified with 2, 4, and 6 g/100 fiber had greater ($P > 0.05$) gel strength than those that were not treated with fiber [66]. The color of the surimi fortified with fiber showed good color properties including slight whitening effect except for 8/100 g of fiber fortification. As demonstrated by differential scanning calorimetry, added fiber did not interfere with the thermal transitions of surimi myosin and actin.

Powdered cellulose is an obvious choice for fortification of surimi with dietary fiber because of its bland flavor and whiteness, as these characteristics do not alter the color and flavor of surimi. They do not interfere with the natural characteristics of the products that are fortified. They instead added whiteness to the color of the surimi product.

7. Conclusions

Superfoods/functional foods can be developed from seafoods including fish protein (ISP) isolate and surimi recovered with isoelectric/precipitation (ISP) with acceptable sensory and nutritional qualities. Fish protein isolates recovered with ISP can serve as a vehicle not just for improving dietary content of ω-3 PUFAs, but the fortification with several useful ingredients. The overall ω-3 PUFAs oil content in fish can be improved by incorporating ω-3 PUFAs oil into the fish, improve the color, texture, rheology of the fish product. It also allows the use of salt substitute to increase the potassium content of the fish whiles reducing sodium content. Interestingly, ISP can be used to recover proteins from different sources including food processing-by products which otherwise would have been difficult to recover using conventional methods. Surimi can be prepared using the conventional water-process, as well as using ISP. It is possible to fortify surimi products with dietary fiber and obtain good color and textural properties. The benefits of recovering fish proteins are numerous, they are reported to contribute to ash content of the product, and thus, fluoride content of the products [43].

References

[1] Milner JA. Functional foods: The US perspective. The American Journal of Clinical Nutrition. 2000;71:1654-1659

[2] Rozin P. The meaning of food in our lives: A cross-cultural perspective on eating and well-being. Journal of Nutrition Education and Behavior. 2005;(37):107-112. DOI: 10.1016/S1499-4046(06)60209-1

[3] Lunn J. Superfoods. Nutrition Bulletin. 2006;31:171-172. DOI: 10.1111/j.1467-3010.2006.00578.x

[4] Siró I, Kápolna E, Kápolna B, Lugasi A. Functional food. Product development, marketing and consumer acceptance — A review. Appetite. 2008;51:456-467. DOI: 10.1016/j.appet.2008.05.060

[5] Thomas PR, Earl RO. Opportunities in the Nutrition and Food Sciences: Research Challenges and the Next Generation of Investigators. Washington, D.C.: National Academies Press; 1994

[6] Tahergorabi R, Matak KE, Jaczynski J, Animal D, Sciences N, Box PO. Fish protein isolate: Development of functional foods with nutraceutical ingredients. Journal of Functional Foods. 2015;**18**:746-756. DOI: 10.1016/j.jff.2014.05.006

[7] Gehring CK, Davenport MP, Jaczynski J. Functional and nutritional quality of protein and lipid recovered from fish processing by-products and underutilized aquatic species using isoelectric solubilization/precipitation. Current Nutrition & Food Science. 2009;**5**:17-39. DOI: 10.2174/157340109787314703

[8] Harnedy PA, Fitzgerald RJ. Bioactive proteins, peptides, and amino acids from macroalgae. Journal of Phycology. 2011;**47**:218-232. DOI: 10.1111/j.1529-8817.2011.00969.x

[9] Hamed I, Özogul F, Özogul Y, Regenstein JM. Marine bioactive compounds and their health benefits: A review. Comprehensive Reviews in Food Science and Food Safety. 2015;**14**:446-465. DOI: 10.1111/1541-4337.12136

[10] Chen YC, Nguyen J, Semmens K, Beamer S, Jaczynski J. Enhancement of omega-3 fatty acid content in rainbow trout (*Oncorhynchus mykiss*) fillets. Journal of Food Science. 2006;**71**:383-389. DOI: 10.1111/j.1750-3841.2006.00115.x

[11] Giltay EJ, Gooren LJG, Toorians AWFT, Katan MB, Zock PL. Docosahexaenoic acid concentrations are higher in women than in men because of estrogenic effects. The American Journal of Clinical Nutrition. 2004;**80**:1167-1174

[12] Ganesan B, Brothersen C, McMahon DJ. Fortification of foods with omega-3 polyunsaturated fatty acids. Critical Reviews in Food Science and Nutrition. 2014;**54**:98-114. DOI: 10.1080/10408398.2011.578221

[13] Kristinsson HG, Rasco BA. Fish protein hydrolysates: Production, biochemical, and functional properties. Critical Reviews in Food Science and Nutrition. 2000;**40**:43-81. DOI: 10.1080/10408690091189266

[14] Shaviklo GR, Thorkelsson G, Arason S, Kristinsson HG, Sveinsdottir K. The influence of additives and drying methods on quality attributes of fish protein powder made from saithe (*Pollachius virens*). Journal of the Science of Food and Agriculture. 2010;**90**:2133-2143. DOI: 10.1002/jsfa.4062

[15] Shaviklo AR. Development of fish protein powder as an ingredient for food applications: A review. Journal of Food Science and Technology. 2013;**52**:648-661. DOI: 10.1007/s13197-013-1042-7

[16] Tina N, Nurul H, Ruzita A. Surimi-like material: Challenges and prospects. International Food Research Journal. 2010;**17**:509-517

[17] Sanmartín E, Arboleya JC, Villamiel M, Moreno FJ. Recent advances in the recovery and improvement of functional proteins from fish processing by-products: Use of protein

glycation as an alternative method. Comprehensive Reviews in Food Science and Food Safety. 2009;**8**:332-344. DOI: 10.1111/j.1541-4337.2009.00083.x

[18] Hultin HO, Kelleher SD. Process for Isolating a Protein Composition from a Muscle Source and Protein Composition. 6,288,216 B1; 1999

[19] Gehring CK, Gigliotti JC, Moritz JS, Tou JC, Jaczynski J. Functional and nutritional characteristics of proteins and lipids recovered by isoelectric processing of fish by-products and low-value fish: A review. Food Chemistry. 2011;**124**:422-431. DOI: 10.1016/j.foodchem.2010.06.078

[20] Choi YJ, Park JW. Acid-aided protein recovery from enzyme-rich Pacific whiting. Journal of Food Science. 2002;**67**:2962-2967. DOI: 10.1111/j.1365-2621.2002.tb08846.x

[21] Kim YS, Park JW, Choi YJ. New approaches for the effective recovery of fish proteins and their physicochemical characteristics. Fisheries Science. 2003;**69**:1231-1239. DOI: 10.1111/j.0919-9268.2003.00750.x

[22] Mireles DeWitt CA, Nabors RL, Kleinholz CW. Pilot plant scale production of protein from catfish treated by acid solubilization/isoelectric precipitation. Journal of Food Science. 2007;**72**. DOI: 10.1111/j.1750-3841.2007.00407.x

[23] Kristinsson HG, Hultin HO. Changes in conformation and subunit assembly of cod myosin at low and high pH and after subsequent refolding. Journal of Agricultural and Food Chemistry. 2003;**51**:7187-7196. DOI: 10.1021/jf026193m

[24] Chen YC, Tou JC, Jaczynski J. Amino acid, fatty acid, and mineral profiles of materials recovered from rainbow trout (*Oncorhynchus mykiss*) processing by-products using isoelectric solubilization/precipitation. Journal of Food Science. 2007;**72**. DOI: 10.1111/j.1750-3841.2007.00522.x

[25] Lansdowne LR, Beamer S, Jaczynski J, Matak KE. Survival of *Escherichia coli* after isoelectric solubilization and precipitation of fish protein. Journal of Food Protection. 2009;**72**:1398-1403

[26] Nolsoe H, Undeland I. The acid and alkaline solubilization process for the isolation of muscle proteins: State of the art. Food and Bioprocess Technology. 2009;**2**

[27] Chen YC, Jaczynski J. Gelation of protein recovered from whole antarctic krill (*Euphausia superba*) by isoelectric solubilization/precipitation as affected by functional additives. Journal of Agricultural and Food Chemistry. 2007;**55**:1814-1822. DOI: 10.1021/jf0629944

[28] Taskaya L, Chen YIC, Beamer S, Tou JC, Jaczynski J. Compositional characteristics of materials recovered from whole gutted silver carp (*Hypophthalmichthys molitrix*) using isoelectric solubilization/precipitation. Journal of Agricultural and Food Chemistry. 2009;**57**:4259-4266. DOI: 10.1021/jf803974q

[29] Chen Y-C, Tou J, Jaczynski J. Amino acid and mineral composition of protein and other components and their recovery yields from whole antarctic krill (Euphausia superba) using isoelectric solubilization/precipitation Introduction. Journal of Food Science. 2009;**74**:H31-H39. DOI: 10.1111/j.1750-3841.2008.01026.x

[30] Bridges KM, Gigliotti JC, Altman S, Jaczynski J, Tou JC. Determination of digestibility, tissue deposition, and metabolism of the omega-3 fatty acid content of krill protein concentrate in growing rats. Journal of Agricultural and Food Chemistry. 2010;**58**:2830-2837. DOI: 10.1021/jf9036408

[31] Lee HJ, Park SH, Yoon IS, Lee G-W, Kim YJ, Kim J-S, et al. Chemical composition of protein concentrate prepared from Yellowfin tuna *Thunnus albacares* roe by cook-dried process. Fisheries and Aquatic Sciences. 2016;**19**:12. DOI: 10.1186/s41240-016-0012-1

[32] Arts MT, Brett MT, Kainz M JTA-TT. Lipids in Aquatic Ecosystems; 2009

[33] Calder PC, Yaqoob P. Omega-3 polyunsaturated fatty acids and human health outcomes. BioFactors. 2009;**35**:266-272. DOI: 10.1002/biof.42

[34] Rizliya V, Mendis E. Biological, physical, and chemical properties of fish oil and industrial applications BT - seafood processing by-products: Trends and applications. In: Kim S-K, editor. Seaf. Process. By-Products Trends Appl. New York, NY: Springer New York; 2014. pp. 285-313. DOI: 10.1007/978-1-4614-9590-1_14

[35] Ungerfeld R, Dago AL, Rubianes E, Forsberg M. Response of anestrous ewes to the ram effect after follicular wave synchronization with a single dose of estradiol-17beta. Reproduction, Nutrition, Development. 2004;**44**:89-98. DOI: 10.1051/rnd

[36] Kris-Etherton PM, Harris WS, Appel LJ. Fish consumption, fish oil, omega-3 fatty acids, and cardiovascular disease. Circulation. 2002;**106**:2747-2757. DOI: 10.1161/01.CIR.0000038493.65177.94

[37] Eaton SB, Konner M. Paleolithic nutrition. A consideration of its nature and current implications. The New England Journal of Medicine. 1985;**312**:283-289. DOI: 10.1056/NEJM198501313120505

[38] Eaton SB, Konner M, Shostak M. Stone agers in the fast lane: Chronic degenerative diseases in evolutionary perspective. The American Journal of Medicine. 1988;**84**:739-749

[39] Simopoulos AP. Omega-3 fatty acids in health and disease and in growth and development. The American Journal of Clinical Nutrition. 1991;**54**:438-463

[40] Tahergorabi R, Beamer SK, Matak KE, Jaczynski J. Chemical properties of ω-3 fortified gels made of protein isolate recovered with isoelectric solubilisation/precipitation from whole fish. Food Chemistry. 2013;**139**:777-785. DOI: 10.1016/j.foodchem.2013.01.077

[41] Kassis NM, Beamer SK, Matak KE, Tou JC, Jaczynski J. Nutritional composition of novel nutraceutical egg products developed with omega-3-rich oils. LWT- Food Science and Technology. 2010;**43**:1204-1212. DOI: https://doi.org/10.1016/j.lwt.2010.04.006

[42] Kassis NM, Gigliotti JC, Beamer SK, Tou JC, Jaczynski J. Characterization of lipids and antioxidant capacity of novel nutraceutical egg products developed with omega-3-rich oils. Journal of the Science of Food and Agriculture. 2012;**92**:66-73. DOI: 10.1002/jsfa.4542

[43] Gigliotti JC, Davenport MP, Beamer SK, Tou JC, Jaczynski J. Extraction and characterisation of lipids from antarctic krill (*Euphausia superba*). Food Chemistry. 2011;**125**:1028-1036. DOI: 10.1016/j.foodchem.2010.10.013

[44] Simopoulos AP. Evolutionary aspects of omega-3 fatty acids in the food supply. Prostaglandins, Leukotrienes, and Essential Fatty Acids. 1999;**60**:421-429

[45] Taskaya L, Chen Y, Jaczynski J. Texture and colour properties of proteins recovered from whole gutted silver carp (*Hypophthalmichthys molitrix*) using isoelectric solubilisation/precipitation. Journal of the Science of Food and Agriculture. 2009;**89**:349-358. DOI: 10.1002/jsfa.3461

[46] Taskaya L, Chen Y, Jaczynski J. Color improvement by titanium dioxide and its effect on gelation and texture of proteins recovered from whole fish using isoelectric solubilization/precipitation. LWT- Food Science and Technology. 2010;**43**:401-408. DOI: 10.1016/j.lwt.2009.08.021

[47] Tahergorabi R, Beamer SK, Matak KE, Jaczynski J. Effect of isoelectric solubilization/precipitation and titanium dioxide on whitening and texture of proteins recovered from dark chicken-meat processing by-products. LWT- Food Science and Technology. 2011;**44**:896-903. DOI: 10.1016/j.lwt.2010.10.018

[48] Feng X, Zhang L, Chen J, Guo Y, Zhang H, Jia C. Preparation and characterization of novel nanocomposite films formed from silk fibroin and nano-TiO_2. International Journal of Biological Macromolecules. 2007;**40**:105-111. DOI: 10.1016/j.ijbiomac.2006.06.011

[49] Tahergorabi R, Beamer SK, Matak KE, Jaczynski J. Functional food products made from fish protein isolate recovered with isoelectric solubilization/precipitation. LWT- Food Science and Technology. 2012;**48**:89-95. DOI: 10.1016/j.lwt.2012.02.018

[50] Pérez-Mateos M, Boyd L, Lanier T. Stability of omega-3 fatty acids in fortified surimi seafoods during chilled storage. Journal of Agricultural and Food Chemistry. 2004;**52**:7944-7949. DOI: 10.1021/jf049656s

[51] Park JW. Ingredient technology for surimi and surimi seafood. In: Park JW, editor. Surimi Surimi Seaf. Boca Raton, FL: Taylor & Francis; 2005, pp. 649-707

[52] Kristinsson GH, Theodore EA, Demir N, Ingadottir B. A comparative study between acid- and alkali-aided processing and surimi processing for the recovery of proteins. Journal of Food Science. 2005;**70**:298-306

[53] Hagen Ø, Solberg C, Sirnes E, Johnston IA. Biochemical and structural factors contributing to seasonal variation in the texture of farmed Atlantic halibut (*Hippoglossus hippoglossus* L.) flesh. Journal of Agricultural and Food Chemistry. 2007;**55**:5803-5808. DOI: 10.1021/jf063614h

[54] Tahergorabi R, Sivanandan L, Beamer SK, Matak KE, Jaczynski J. A three-prong strategy to develop functional food using protein isolates recovered from chicken processing by-products with isoelectric solubilization/precipitation. Journal of the Science of Food and Agriculture. 2012;**92**:2534-2542. DOI: 10.1002/jsfa.5668

[55] Szczesniak AS. Texture is a sensory property. Food Quality and Preference. 2002;**13**:215-225. DOI: 10.1016/S0950-3293(01)00039-8

[56] Geleijnse JM, Witteman JC, Bak AA, den Breeijen JH, Grobbee DE. Reduction in blood pressure with a low sodium, high potassium, high magnesium salt in older subjects with mild to moderate hypertension. BMJ 1994;**309**:436-440

[57] Jaczynski J, Park JW. Physicochemical changes in Alaska pollock surimi and surimi gel as affected by electron beam. Journal of Food Science. 2004;**69**:C53-C57. DOI: 10.1111/j.1365-2621.2004.tb17855.x

[58] Anderson JW, Baird P, Davis RHJ, Ferreri S, Knudtson M, Koraym A, et al. Health benefits of dietary fiber. Nutrition Reviews. 2009;**67**:188-205. DOI: 10.1111/j.1753-4887.2009.00189.x

[59] InterAct C. Dietary fibre and incidence of type 2 diabetes in eight European countries: The EPIC-InterAct Study and a meta-analysis of prospective studies. Diabetologia. 2015;**58**:1394-408

[60] Stephen AM, Champ MM-J, Cloran SJ, Fleith M, van Lieshout L, Mejborn H, et al. Dietary fibre in Europe: Current state of knowledge on definitions, sources, recommendations, intakes and relationships to health. Nutrition Research Reviews. 2017;**30**:2830-2837. DOI: 10.1017/S095442241700004X

[61] Sofi S, Singh J, Rafiq S, Rashid R. Fortification of dietary fiber ingriedents in meat application: A review. International Journal of Biochemistry Research & Review. 2017;**19**:1-14. DOI: 10.9734/IJBCRR/2017/36561

[62] Sánchez-Alonso I, Solas MT, Borderías AJ. Technological implications of addition of wheat dietary fibre to giant squid (*Dosidicus gigas*) surimi gels. Journal of Food Engineering. 2007;**81**:404-411. DOI: https://doi.org/10.1016/j.jfoodeng.2006.11.015

[63] Cardoso C, Mendes R, Vaz-Pires P, Nunes ML. Effect of dietary fibre and MTGase on the quality of mackerel surimi gels. Journal of the Science of Food and Agriculture. 2009;**89**:1648-1658. DOI: 10.1002/jsfa.3636

[64] Cardoso C, Mendes R, Nunes ML. Effect of transglutaminase and carrageenan on restructured fish products containing dietary fibres. International Journal of Food Science and Technology. 2007;**42**:1257-1264. DOI: 10.1111/j.1365-2621.2006.01231.x

[65] Sánchez-Alonso I, Haji-Maleki R, Borderias AJ. Wheat fiber as a functional ingredient in restructured fish products. Food Chemistry. 2007;**100**:1037-1043. DOI: 10.1016/j.foodchem.2005.09.090

[66] Debusca A, Tahergorabi R, Beamer SK, Matak KE, Jaczynski J. Physicochemical properties of surimi gels fortified with dietary fiber. Food Chemistry. 2014;**148**:70-76. DOI: 10.1016/j.foodchem.2013.10.010

Chapter 4

Health Potential for Beer Brewing Byproducts

Abstract

The beer brewing process involves malting, milling, mashing, boiling, cooling, and fermentation. Approximately 20 kg of byproducts are produced for every 100 liters of beer brewed, of which brewer's spent grains (SG), spent hops (SP), and surplus yeasts (SY) account for approximately 85, 5, and 10%, respectively. SG is rich in cellulose, protein, essential amino acids, phenolics and mineral; SP is rich in nitrogen free extract, fiber and protein; SY is rich in proteins and saccharides; where both SP and SY also are rich in prenylflavonoids and hop bitter acids. Although several nutrients or functional components have been found in such beer brewing byproducts, most of these byproducts are used as animal feed and fertilizers since insufficient research has been devoted to the physiological activities for human. To date, only activities of antiobesity and antiproliferation of cancer cells were possessed by SY. Hence, further research is required to clarify the health potential and novel application of these byproducts for environmental protection and other economic activities.

Keywords: beer brewing byproducts, spent grains, spent hops, surplus yeasts, health potential

1. Introduction

Beer is a popular alcoholic drink around the world. Beer is mainly made from sprouted cereals (mostly barley malt) that are fermented with *Saccharomyces* after hops and water have been added. The product, beer, contains natural carbon dioxide and beer yeast and has characteristics including lasting foam, low alcohol content, and richness of nutrients; because of these features, beer is sometimes called "liquid bread."

Although studies have shown that beer contains several functional components, amounts of these functional compounds in beer are diluted since high water content of beer, and the substantial calorie value and alcoholic content also limit health effects of beer. Conversely, brewing byproducts may contain more functional components during separation procedure. Their further recycling not only achieves waste reduction but also increases their commercial value. This review is aimed at providing detailed information on beer brewing process and the related byproduct generated, as well as the content of functional components in beer brewing byproducts and their health potential.

2. Brewing beer and brewing byproducts

The common beer brewing process contains four major stages: malting, mashing, wort boiling, and fermentation [1]. Related processes and the corresponding byproducts are shown in **Figure 1** [2].

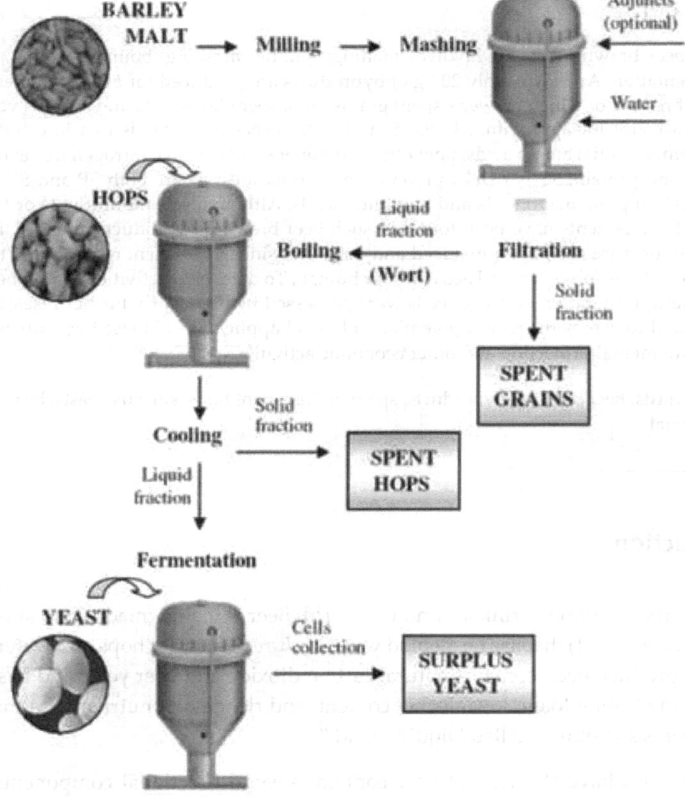

Figure 1. Schematic representation of the brewing process and points where the main byproducts are generated Mussatto [2].

After barley grain has been sieved to remove thinner barley, dust, and impurities, it is steeped in water for germination. During this malting stage, enzymes decompose the complex protein–carbohydrate structure into smaller molecules and expose the starch granules inside the endosperm. Barley grain is allowed to grow only partially because fully grown consumes excessive amounts of starch, and this affects the subsequent fermentation process. Therefore, after the barley grain has germinated, it is dried to halt growth [3].

The malted barley from the malting stage is milled to crush the endosperm completely. Next, the crushed barley is mixed with water and the temperature of the mixture is slowly increased from 37 to 78°C to enhance the hydrolysis capacity of α-amylase and β-amylase. This process converts starch into fermentable sugars (mainly maltose and maltotriose) and unfermentable sugars (maltodextrin), whereas protein is degraded into peptide and amino acids. The process of starch conversion with enzymes is called "mashing". Subsequently, the mixture is filtered; the resultant liquid part is called "wort" and the solid residue is called "spent grains" [4]. During processing, hot water is sometimes added to spent grain to collect second wort. Most beer brews are fermented with a mixture of first and second wort.

During wort boiling step, hops (*Humulus lupulus* L.) are added to the wort, which is then boiled for approximately 1 hour to sterilize the wort and deactivate the glycolysis enzymes. During this process, the components in hops that create a fragrant and the bitter flavor enter the wort and provide the beer with its unique bitter flavor [5]. Moreover, hops enhance beer quality, stabilize bubbles, inhibit glycolysis, and preserve and clarify the wort [6]. After wort boiling has been completed, the wort is filtered again. The solid matter obtained is called "spent hops", which is the second byproduct of the brewing process.

The wort is cooled to between 12 and 18°C for yeast inoculation. Commonly employed yeasts include *Saccharomyces bayanus*, *S. cariocanus*, *S. cerevisiae*, *S. kudriavzevii*, *S. mikatae*, and *S. paradoxus* [1]. During fermentation, the amount of yeast multiplies by three- to six-fold, converting fermentable sugar into ethanol and carbon dioxide; the carbon dioxide is dissolved into the beer. Inoculating different species of yeast gives beer different flavors. The color of beer gradually becomes transparent amber [7]. After the beer aging process has been completed, the yeast cell biomass is collected with all other insoluble materials settled at the tank bottom during filtration. Those insoluble materials are the third brewing byproduct, namely surplus yeasts [1].

3. Composition and current applications of beer brewing byproducts

The beer brewing process produces three byproducts (i.e., spent grains, spent hops, and surplus yeasts. These byproducts remain limited in terms of application and are often used in animal feeds in animal husbandry [2]. The following subsection briefly introduces spent grains, spent hops, and surplus yeasts.

3.1. Spent grains

Spent grains, the most abundant byproduct of beer brewing, accounting for approximately 85% of a brew's total byproduct yield. Spent grains are mainly composed of barley shells, remaining endosperm starch granules, and other cereal additives (e.g., wheat, rice, and corn) added to give beer its unique flavor [1]. The main chemical compositions of spent grains include cellulose, hemicellulose, and lignin. Cellulose and hemicellulose account for 50% of all spent grains. Moreover, spent grains are rich in protein, essential amino acids, minerals, and single sugars (glucose, xylose, and arabinose) [8, 9].

Spent grains are also abundant in phenolics such as ferulic acid, *p*-coumaric acid, syringic acid, vanillic acid and *p*-hydroxybenzoic acid [10]. Mussatto [2] stated that in contrast to that of barley, the composition of spent grains is affected by the following conditions: the barley type, the time of year when barley is planted, the degree to which barley is grinded, and the conditions for malting and glycolysis.

Because spent grains are rich in protein, cellulose, and minerals, they are often used to feed ruminants [11]. Studies have investigated adding spent grains to bread and snacks to increase their fiber content [12]. Spent grains contain many functional groups such as hydroxyl, amine, and carboxyl, all of which can be used as biosorbents. For examples, spent grains can effectively remove volatile substances in exhaust gas and remove contaminants in waste water such as heavy metals (lead, chromium, zinc, copper, and cadmium) and dyes [1, 9, 13]. Moreover, spent grains have proven effective in antioxidation and for antibacterial and anti-inflammatory uses, and can even serve as a precursor for food flavors [14], a medium for producing microbes and enzymes [15], and a raw material for bioethanol [11]. Compared with spent hops and surplus yeasts, spent grains can more easily be reused. In addition, because of the advantages of being high-yielding and containing complex chemical components, spent grain can be used in a variety of areas [4].

3.2. Spent hops

Hops (*Humulus lupulus* L.) are a perennial herbaceous and dioecious plant grown mainly in Europe and North America. The main component of dried hops is fiber, followed by hop bitter acids and protein. Dried hops also contain ash, salts, polyphenols, tannins, and oil [16]. The fiber in hops is composed of xylose, mannose, galactose, and glucose. Of these, glucose and xylose are the most prevalent. In addition, hops contain pectin, uronic acid, rhamnose, and arabinose [17].

Of the added hops, only 15% remain in the final product of beer; the other 85% are residue in spent hops. Therefore, spent hops are abundant in nitrogen-free extract, fiber, and protein [4] and rich in essential amino acids; the composition of spent hops is similar to that of hops [18].

Although spent hops are rich in nitrogen, carbon, and protein, their bitterness limits their application in food. The bitterness of spent hops can be removed with *Candida parapsilosis* fermentation; after this removal, spent hops can be made into animal feed supplements [19]. Because spent hops are rich in nitrogen, they can be used as soil improver and fertilizer.

Spent hops have other applications; for example, oxidized or hydrolyzed spent hops to obtain products with commercial value, including spices, carbohydrates, and organic acids [20]. Furthermore, adding spent hops during the beer brewing process increases yeast activity, which in turn increases the beer yield [21].

3.3. Surplus yeasts

Surplus yeasts are the second most abundant byproduct of beer brewing, accounting for approximately 10% of the total amount of byproducts [1]. After being separated from beer, surplus yeasts are first heated to halt its activity [22]. The composition of *Saccharomyces cerevisiae* determines the composition of surplus yeasts. The major element in yeast is carbon, which accounts for up to 50% of the dry weight, followed by oxygen (30–35%), nitrogen (5%), helium (5%) and phosphorus (1%). Therefore, surplus yeasts are mainly composed of proteins and saccharides [2]. Briggs et al. [5] indicated that because cells vary in terms of their physiological conditions and numbers of bacterial growth cycles, the compositions of different types of surplus yeast also vary.

The protein compositions in surplus yeasts, spent grains, and spent hops are similar in that they all contain several essential amino acids, vitamins, and minerals [2]. The total mineral content in yeast is 5–10% of their dry weight; yeast is especially high in potassium and phosphorus [2].

Chae et al. [23] stated that compared with spent hops, surplus yeast contains more proteins, vitamins, and amino acids, and thus is more often used in animal feeds and nutritional supplements. However, because yeast contains ribonucleic acid (RNA), which is produced by metabolizing uric acid, eating excessive amounts of surplus yeasts can cause gout. Consequently, yeast is limited in the extent to which it can be used in foods [22]. Surplus yeast is often used as a source of carbon and nitrogen when cultivating microorganisms and in health foods in different food seasons [24]. β-glucan, a type of hydrocolloid, can be extracted from surplus yeasts for its ability to improve food characteristics as a thickener, water retaining agent, oil-retaining agent, emulsifier, or foam stabilizer [25]. Parvathi et al. [26] discovered that surplus yeast has the ability to absorb lead.

4. Functional components in beer brewing byproducts

4.1. Dietary fibers

Study revealed that spent grains contained 60–68% of insoluble dietary fibers (IDF) [27]. Another study revealed that dried hops contained 40–50% IDF [16], and beer contained 0.4–6.2 g/L dietary fibers [28]. Lin [29] reported that spent grains contained the most IDFs, followed by hops, spent hops, and surplus yeasts. Spent hops had the highest soluble dietary fibers, followed by spent grains, spent hops, and surplus yeasts. The aforementioned results indicated that during the beer brewing process, dietary fibers in malt and hops were left in spent grains, spent hops, and surplus yeasts, whereas few of them are retained in beer.

4.2. Vitamins

Lewis and Young [30] stated surplus yeasts contained vitamins B1, B2, B3, B5, B6, and B9 in proportions of 15, 7, 50, 10, 3, and 4 mg/100 g, respectively. Bamforth [31] discovered that beer contained vitamin B1 (0.003–0.08 mg/L), B2 (0.02–0.8 mg/L), B3 (3–8 mg/L), B6 (0.07–1.7 mg/L), and B12 (0.003–0.03 mg/L), whereas vitamin E was not detected. However, Lin [29] reported that neither vitamins B, C, nor E were detected in beer brewing byproducts, possibly because the boiling stage in the beer brewing process causes splitting decomposition in vitamins due to high heat and oxidation. Moreover, spent hops and surplus yeast are considered food waste, and thus if brewers do not store them properly, vitamins can be lost.

4.3. Phenolic compounds

Studies have revealed that the sources of phenolic compounds in beer were barley and hops, with barley providing 70–80% and hops providing 20–30% [2]. During the boiling, filtering, and ageing processes, the amount of phenolic compounds changed because phenolic compounds were used to scavenge free radicals and stabilize the flavoring materials and foam in beer [32]. During the cooling process after hops were added, some insoluble materials and polyphenols in the hops formed complices with protein in the wort, precipitated, and were then filtered out together in the next stage, resulting in phenolic compounds being present in beer brewing byproducts.

4.3.1. Phenolic acids

Spent grains contains ferulic acid, *p*-coumaric acid, caffeic acid, syringic acid vanillic acid, and *p*-hydroxybenzoic acid. Among these, one study observed that the content of ferulic acid was the highest [10, 33].

4.3.2. Prenylflavonoids

Prenylflavonoids are a subclass of flavonoids with a prenyl group of aromatic rings. Based on whether they have an open ring structure, they are divided into prenylchalcones and prenylflavanones (**Figure 2**) [34]. If prenylchalcones have a methoxy group attached to C6 in the aromatic ring, they are xanthohumol, whereas if they have a hydroxyl group attached, they are desmethylxanthohumol. If prenylflavanones have a methoxy group attached to C5 in the aromatic ring, they are isoxanthohumol, whereas if they have a hydroxyl attached, they are 8-prenylnaringenin.

To date, related studies on prenylflavonoids have mainly focused on prenylflavonoids in relation to beer and hops. Most prenylchalcones in hops and beer were xanthohumol and dexmethylxanthohumol, whereas most prenylflavanones were isoxanthohumol, 6-prenylnaringenin, and 8-prenylnaringenin [35].

Regarding byproducts, Kao and Wu [36] indicated that beer lees contained isoxanthohumol (36.22 μg/g), xanthohumol (7.84 μg/g), 8-prenylnaringenin (19.17 μg/g), and 6-prenylnaringenin (29.56 μg/g). Lin [29] reported that surplus yeasts were most abundant in isoxanthohumol, followed by xanthohumol, 6-prenylnaringenin, and 8-prenylnaringenin. Spent hops contained mainly xanthohumol, followed by isoxanthohumol and 6-prenylnaringenin.

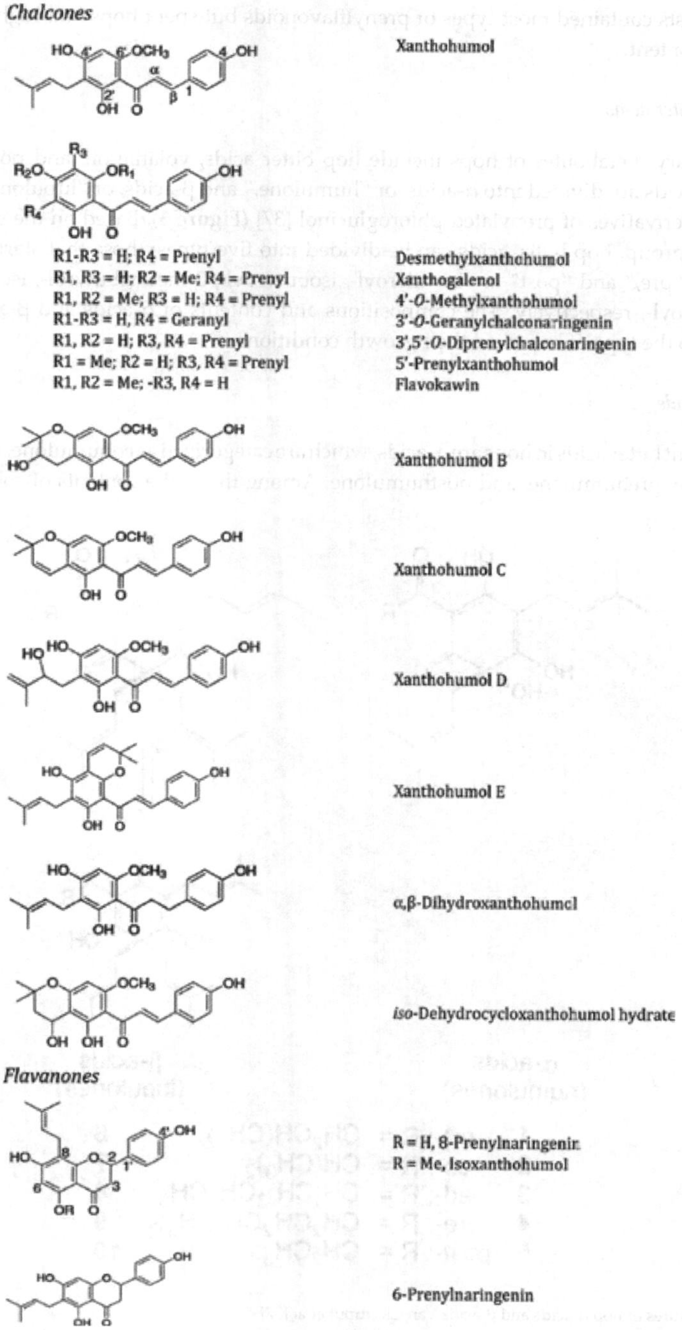

Chalcones

Xanthohumol

R1-R3 = H; R4 = Prenyl	**Desmethylxanthohumol**
R1, R3 = H; R2 = Me; R4 = Prenyl	**Xanthogalenol**
R1, R2 = Me; R3 = H; R4 = Prenyl	**4'-*O*-Methylxanthohumol**
R1-R3 = H, R4 = Geranyl	**3'-*O*-Geranylchalconaringenin**
R1, R2 = H; R3, R4 = Prenyl	**3',5'-*O*-Diprenylchalconaringenin**
R1 = Me; R2 = H; R3, R4 = Prenyl	**5'-Prenylxanthohumol**
R1, R2 = Me; -R3, R4 = H	**Flavokawin**

Xanthohumol B

Xanthohumol C

Xanthohumol D

Xanthohumol E

α,β-Dihydroxanthohumol

***iso*-Dehydrocycloxanthohumol hydrate**

Flavanones

R = H, 8-Prenylnaringenin
R = Me, Isoxanthohumol

6-Prenylnaringenin

Figure 2. Prenylated chalcones and flavanones from hops Stevens and Page [34].

Surplus yeasts contained most types of prenylflavonoids but spent hops had higher prenylflavonoid content.

4.3.3. Hop bitter acids

The secondary metabolites of hops include hop bitter acids, volatile oil, and polyphenols. Hop bitter acids are divided into α-acids, or "humulone," and β-acids, or "lupulone," both of which are derivatives of prenylated phloroglucinol [37] (**Figure 3**). Based on the side chains on the acyl group, hop bitter acids can be divided into five types; those that start with "n," "co," "ad," "pre," and "post" are isovaleroyl-, isobutyroyl-, 2-methylbutyrol-, isohexanoyl-, and propanoyl-, respectively. The compositions and contents of α-acids and β-acids differ according to the type of hops and their growth conditions.

4.3.3.1. α-Acids

The dominant bitter acids in hops are α-acids, which are categorized as cohumulone, humulone, adhumulone, prehumulone, and posthumulone. Among these, the contents of cohumulone,

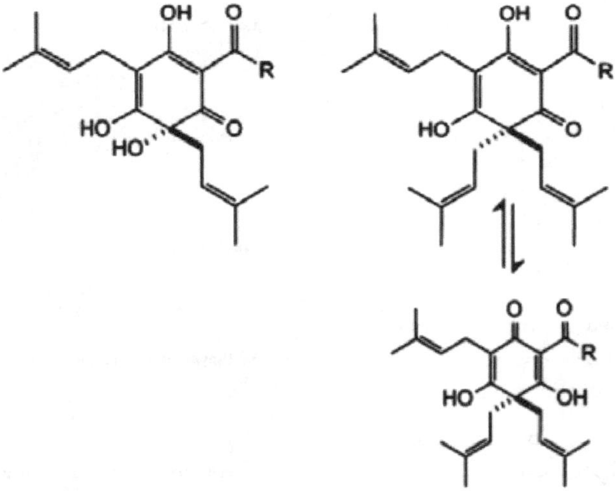

α-acids (humulones)

β-acids (lupulones)

1 (n)-: R = $CH_2CH(CH_3)_2$ 6
2 co-: R = $CH(CH_3)_2$ 7
3 ad-: R = $CH(CH)_3CH_2CH_3$ 8
4 pre-: R = $CH_2CH_2CH(CH_3)_2$ 9
5 post-: R = CH_2CH_3 10

Figure 3. Structures of hop α-acids and β-acids Van Cleemput et al [37].

humulone, and adhumulone were higher, accounting for 35–70%, 20–65%, and 10–15% of α-acids, respectively [38]. Because α-acids are less water soluble and because during wort boiling, α-acids isomerize into iso-α-acids that are more water soluble, only approximately 25% of α-acids are retained in beer [6].

4.3.3.2. β-Acids

β-acids mainly exists in hops in the forms of lupulone and colupulone, both of which accounted for20–55% of the total β-acid content; adlupulone accounted for 10–15% and the contents of prelupulone and postlupulone were very low [37].

Because the tertiary alcohol on the C6 in β-acids is replaced by a prenyl side chain, its acidity is lower than that in α-acids. β-acids are highly sensitive to oxygen; they can easily autoxidize into stable hulupones, which include cohulupone, hulupone, and adhulupone [39]. Hulupones exist in matured hops and beer at degrees of condensation of 2–10 ppm, and provide beer with a bitter taste. Hulupones degrade into hulupinic acid, which is not bitter. The content of β-acids in beer is lower than that of α-acids, and thus β-acids do not considerably affect beer quality [40].

4.3.3.3. Iso-α-acids

During the wort boiling process, α-acids isomerize through the acyloin-type ring into iso-α-acids. The condensation of iso-α-acids in beer is a low 15–100 ppm; however, these acids are the source of 80% of the bitter taste in beer [41].

When heating α-acids in an alkaline solution containing divalent cations such as magnesium ion or calcium ion as a catalyst for isomerization, α-acids become *cis*-isomers and *trans*-isomers at a ratio of 1 to 1 [42]. When iso-α-acids are placed under ultraviolet and visible light (350–500 nm) with photosensitizers such as riboflavin, photo-oxidation occurs to break the side chain on C4, forming dehydrohumulinic acid and 3-methyl-2-butene-1-thiol, which is also known as "skunky thiol" because it is the main source of unfavorable flavors in beer. In addition, through reduction, iso-α-acids create *trans*-dihydro-isohumulones, and also *trans*-tetrahydro-isohumulones and *trans*-hexahydro-isohumulones, both of which can stabilize the foam produced by beer [37].

4.3.3.4. Hop bitter acids in brewing byproducts

Kao and Wu [36] discovered cohumulone, humulone, adhumulone, colupulone, lupulone], and adlupulone in beer lees. Among these components, humulone was the most dominant. Lin [29] reported findings of 8017 and 1130 μg/g of hop bitter acids in spent hops and surplus yeast, respectively. The hop bitter acids detected in spent hops were cohulupone, hulupone, adhulupone, *trans*-isocohumulone, *trans*-isohumulone, *trans*-isoadhumulone, cohumulone, humulone, adhumulone, prehumulone, adprehumulone, colupulone, lupulone, adlupulone, prelupulone, and adprelupulone. The same bitter acids were detected in surplus yeasts, except for *trans*-isocohumulone, prelupulone, and adprelupulone. In spent hops, colupulone had the highest content, with lupulone second. In surplus yeasts, humulone had the highest content, followed by colupulone.

5. Health benefits and potential of byproducts from beer brewing

To date, insufficient research has been devoted to the physiological functions of byproducts from beer brewing. However, these byproducts are rich in prenylflavonoids and hop bitter acids, which, as proven by many studies, have several health benefits that are detailed in this section. Therefore, beer brewing byproducts may have specific physiological activities. This section organizes physiological activities observed in previous studies, however, further research is required to clarify the physiological functions of the byproducts of beer brewing.

5.1. Antiobesity

Surplus yeast hinders the survival ratios of preadipocytes and adipocytes and can reduce cell apoptosis. The surplus yeast extract process arrests preadipocytes at G2/M, and cyclin B1 (a protein related to the cell cycle) substantially increases in number. Adding surplus yeast extract during the process of preadipocyte differentiation into adipocytes effectively reduces the number of adipocytes and hinders the formation of triglycerides [43].

Animal models have proven that feeding of brewer's yeast biomass with ethanol extract to male Sprague-Dawley rats could reduce the weight of fat around the kidneys and paraplegia, reduced triglyceride content in serum and the liver, and increased antioxidant capacity in the liver [44].

When xanthohumol and isoxanthohumol treated preadipocyte 3T3-L1, the expressions of transcription factors for lipid metabolism reduced, hence preventing fat tissue from forming and inducing apoptosis of adipocytes [45]. Xanthohumol could suppress the weight of rats with high-fat diets, reduce weight gain around the liver, and reduce the triglyceride content in their serum and liver [46].

5.2. Antiproliferation of cancer cells

Surplus yeasts extract in non-small cell lung carcinoma A549 increased the protein expression from those of the mitogen-activated protein kinase family, including p-ERK1/2, p-JNK, and p-p38, and suppressed the expression of cyclin E1 and arrested the cell cycle at G0/G1, thereby reducing the survival of A549 cells. In H460 lung cancer cells, surplus yeast extract significantly increased the expressions of proteins p-ERK1/2 and p-p38 and decreased the protein content of cyclin D1 and cyclin E1, thereby arresting cell cycle at G0/G1. The main suppressants in these two experiments were xanthohumol and hop bitter acids [47].

5.3. Anti-inflammatory activity

Studies have shown that xanthohumol, β-acids and hexahydro-β-acids could suppress the generation of chemokine and cytokine, as well as their gene expression in inflammatory cells [48, 49]. In animal models, xanthohumol was considered effective in reducing liver inflammation, thereby minimizing liver fibrosis [50].

Hop bitter acids could suppress the activation of nuclear factor-κB (NF-κB) in liver astrocytes, and could also suppress the generation of monocyte chemoattractant protein-1 (MCP-1) and α-smooth muscle actin (α-SMA) to achieve anti-inflammatory and antifibrotic outcomes [51]. In an animal test, feeding 1.25 g of hop extract to LPS-induced mice for 10 days reduced the PGE2 content in their blood [52].

5.4. Antioxidant activity

Feeding xanthohumol to rats with chemical hepatitis could increase the glutathione content and the activities of antioxidant enzymes in liver [53]. Dorn et al. [54] reported that after xanthohumol treatment, the active oxygen content due to ischemia–reperfusion injury had reduced. In addition, isoxanthohumol was able to remove active oxygen [55].

Iso-α-acids, α-acids, and β-acids all have the abilities to clear the peroxynitrite free radical and reduce lipid peroxidation [56]. Namikoshi et al. [57] reported that the rat kidneys on a high-salt diet produced excessive amounts of active oxygen, resulting in kidney tissue damage due to oxidation. Iso-α-acids were able to suppress the formation of active oxygen to achieve the goal of antioxidation.

5.5. Estrogen activity

8-prenylnaringenin exhibited a structure similar to that of estrogen and could be bound with estrogen acceptors, indicating that 8-prenylnaringenin also exhibited activity similar to that of estrogen [58]. Christoffel et al. [59] reflected that feeding Sprague Dawley rats 8-prenylnaringenin for 3 months could reduce the content of follicle-stimulating hormone and luteinizing hormone in serum, increase the weight of the uterus and stimulate the secretion of insulin-like growth factor I and prolactin.

5.6. Antiangiogenesis

Xanthohumol could suppress the formation of vascular endothelial growth factor (VEGF) and interleukin-8 in pancreatic cancer cells, and the antiangiogenesis effect was achieved through suppressing NF-κB [60].

Shimamura et al. [61] proved that 100 μM of humulone was effective in suppressing the formation of the VEGF and could prevent angiogenesis from Co26s cells. An amount of 2.5 to 50 μg/mL of lupulone could suppress the proliferation of human umbilical vein endothelial cells and the secretion of fibronectin.

5.7. Anticancer activity

Xanthohumol, isoxanthohumol, 8-prenylnaringenin, and 6-prenylnaringnin were all effective in suppressing the proliferation of human prostate cancer cells PC-3 and DU145 [62]. In addition, xanthohumol was able to reduce gene expressions related to the NOTCH1 pathway, thereby arresting the cell cycle of human ovarian carcinoma cells SKOV3 and OVCAR3 in the G2/M period and further causing apoptosis [63].

Festa et al. [64] revealed that xanthohumol activated caspase-3 and caspase-9, through mitochondrial depolarization, releasing cytochrome C, reducing Bcl-2 protein expression, and increasing the oxidation pressure, causing apoptosis of human glioblastoma multiforme tumor cell (T98G). Xanthohumol could prevent carcinogen metabolites from being processed by phase I enzymes (e.g. cytochrome P450) or activated by cytochrome isomers (e.g., cytochrome 1A1 and cytochrome 1A2). Xanthohumol was also able to promote carcinogens excreted from human body by increase their water-solubility that effected by phase II enzymes (glutathione S-transferase and UDP-glucuronyl transferase) [34].

Apoptosis of HL-60 (human promyelocytic leukemia cells) and SW 620 (human colorectal adenocarcinoma cells) can be induced by β-acids and α-acids [65]. Yasukawa et al. [66] maintained that humulone can suppress tumor formation in mice treated by 7,12-dimethylbenz[a] anthracene.

5.8. Antianxiety activity

In contrast to β-acids, which are sedatives, α-acids are the major anxiolytic substances in hops [67]. Therefore, through *r*-aminobutyric acid A receptors, 5-hydroxytryptamine receptors, and melatonin, hops were able to alter the central nervous system to achieve the effects of sedation and sleep quality improvement [68].

5.9. Metabolic syndrome prevention

Isohumulones could activate peroxisome proliferator activated receptors α (PPARα) to increase sensitivity to insulin and help protect against type II diabetes [69]. Miura et al. [70] revealed that isohumulone could increase the mRNA expressions of acyl-CoA oxidase, acyl-CoA synthetase, the fatty acid transport protein, and lipoprotein lipase and reduce the mRNA activity of Apo CIII to achieve regulation of liver lipid synthesis.

5.10. Antiosteoporosis

Tobe et al. [71] discovered that α-acids were able to suppress dentin bone loss but that cohumulone was unable to exert the suppression effect. The researchers also discovered that β-acids were the most effective acids in suppressing dentin bone loss. Ding et al. [72] have evaluated cytokines such as interleukin-6 and TNF-α in serum and have proven that the cytokine content had a negative correlation with bone mass. Therefore, the effect of hop bitter acids on inflammation could be applied for osteoporosis prevention and treatments.

6. Conclusion

Prenylflavonoids and hop bitter acids are the most important functional components from hop, and they may remain in the byproducts especially spent hops and surplus yeasts during beer brewing. Both prenylflavonoids and hop bitter acids have been confirmed to possess many physiological activities in current research but studies for establishing the proper health benefit of beer brewing byproducts remained uncertain. As mentioned in the preceding section, the future considerations and latent problems have to be emphasized. First, beer brewing technology should be optimized in order to minimize the amounts of waste arising. Second, methods for complete recovery of by-products during beer brewing on a large scale and at an affordable level should be developed. Third, specific analytical methods for the characterization and quantification of organic micronutrients and other functional compounds need to be built. Fourth, the bioactivity, bioavailability and toxicology of the functional components in beer brewing byproducts need to be carefully assessed by *in vitro* and *in vivo* studies.

References

[1] Kumar SR. In: Chandrasekaran M, editor. Valorization of Food Processing By-products. Boca Raton, Florida, USA: CRC Press; 2012

[2] Mussatto SI. Biotechnological potential of brewing industry by-products. In: Singh nee' Nigam P, Pandey A, editors. Biotechnology for Agro-industrial Residues Utilization. Berlin, Germany: Springer; 2009. pp. 313-326

[3] Protz R. The ultimate encyclopedia of beer: The definitive guide to the world's great brews. New York: Smithmark Publishing; 1995

[4] Mussatto SI, Dragone G, Roberto IC. Brewer's spent grain: Generation, characteristics and potential applications. Journal of Cerea Science. 1995;43:1-14. DOI: 10.1016/j.jcs.2005.06.001

[5] Briggs DE, Boulton CA, Brookes PA, Stevens R. Brewing: Science and Practice. Cambridge: Woodhead Publishing; 2004

[6] De Keukeleire DD. Fundamentals of beer and hop chemistry. Quimica Nova. 2000;**23**: 108-112

[7] Hartmeier W, Reiss M. Production of beer and wine. In: Industrial Applications. 2nd ed. Berlin, Germany: Springer; 2010. pp. 59-77

[8] Mussatto SI, Roberto IC. Chemical characterization and liberation of pentose sugars from brewer's spent grain. Journal of Chemical Technology & Biotechnology. 2006;**81**:268-274. DOI: 10.1002/jctb.1374

[9] Adelagun ROA, Itodo AU, Berezi EP, Oko OJ, Kamba EA, Andrew C, Bello HA Adsorptive removal of Cd^{2+} and Zn^{2+} from aqueous system by BSG. Journal of Chemistry and Materials Research. 2014;**6**:104-112

[10] Mussatto SI, Dragone G, Roberto IC. Feruic and *p*-coumaric acids extraction by alkaline hydrolysis of brewer's spent grain. Industrial Crops & Products. 2007;**25**:231-237. DOI: 10.1016/j.indcrop.2006.11.001

[11] Aliyu S, Bala M. Brewer's spent grain: A review of its potentials and applications. African Journal of Biotechnology. 2011;**10**:324-331. DOI:10.5897/AJBx10.006

[12] Stojceska V, Ainsworth P, Plunkett A. The recycling of brewer's processing by-product into ready-to-eat snacks using extrusion technology. Journal of Cereal Science. 2008;**47**:469-479. DOI: 10.1016/j.jcs.2007.05.016

[13] Silva JP, Sousa S, Rodrigues J, Antunes H, Porter JJ, Goncalves I, Ferreira-Dias S. Adsorption of acid orange 7 dye in aqueous solutions by spent brewery grains. Separation and Purification Technology. 2004;**40**:309-315. DOI: 10.1016/j.seppur.2004.03.010

[14] Mussatto SI, Fernandes M, Dragone G. Brewer's spent grain as raw material for lactic acid production by *Lactobacillus delbrueckii*. Biotechnology Letters. 2007;**29**:1973-1976. DOI: 10.1007/s10529-007-9494-3

[15] Xiros C, Christakopouls P. Biotechnological potential of brewers spent grain and its recent applications. Waste and Biomass Valorization. 2012;**3**:213-332. DOI: 10.1007/s12649-012-9108-8

[16] Benitez JL, Forster A, De Keukeleire D, Moir M, Sharpe FR, Verhagen LC, Westwood KT. Hops and Hop Products. Nurnberg, Germany: Verlag Hans Carl; 1997

[17] Oosterveld A, Voragen AGJ, Schols HA. Characterization of hop pectins shows the presence of an arabinogalactan-protein. Carbohydrate Polymers. 2002;**49**:407-413. DOI: 10.1016/S0144-8617(01)00350-2

[18] Wallen SE, Marshall Jr HF. Protein quality evaluation of spent hops. Journal of Agricultural and Food Chemistry. 1979;**27**:635-636. DOI: 10.1021/jf60223a027

[19] Huszcza E, Bartmańska A. The implication of yeast in debittering of spent hops. Enzyme and Microbial Technology. 2008;**42**:421-425. DOI: 10.1016/j.enzmictec.2008.01.004

[20] Laufenberg G, Kunz B, Nystroem M. Transformation of vegetable waste into value added products: (A) the upgrading concept; (B) practical implementation. Bioresource Technology. 2003;**87**:167-198

[21] Fischer K, Bipp HP. Generation of organic acids and monosaccharides by hydrolyticand oxidative transformation of food processing residues. Bioresource Technology. 2005;**96**:831-842. DOI: 10.1016/j.biortech.2004.07.003

[22] Soares EV, Soares HM. Cleanup of industrial effluents containing heavy metals: a new opportunity of valorising the biomass produced by brewing industry. Applied Microbiology and Biotechnology. 2013;**97**:6667-6675. DOI: 10.1007/s00253-013-5063-y

[23] Chae HJ, Joo H, In MJ. Utilization of brewer's yeast cells for the production of food-grade yeast extract. Part I: Effects of different enzymatic treatments on solid and protein recovery and flavor characteristics. Bioresource Technology. 2001;**76**:253-258. DOI: 10.1016/S0960-8524(00)00102-4

[24] Rakin M, Vukasinovic M, Siler-Marinkovic S. Contribution of lactic acid fermentation to improved nutritive quality vegetable juices enriched with brewer's yeast autolysate. Food Chemistry. 2007;**100**:599-602. DOI: 10.1016/j.foodchem.2005.09.077

[25] Thammakiti S, Suphantharika M, Phaesuwan T. Preparation of spent brewer's yeast-glucans for potential applications in the food industry. International Journal of Food Science & Technology. 2004;**39**:21-29. DOI: 10.1111/j.1365-2621.2004.00742.x

[26] Parvathi K, Nagendran R, Nareshkumar R. Lead biosorption onto waste beer yeast by-product, a means to decontaminate effluent generated from battery manufacturing industry. Electronic Journal of Biotechnology. 2007;**10**:92-105. DOI: 10.2225/vol10-issue1-fulltext-13

[27] Russ W, Mörtel H, Meyer-Pittroff R. Application of spent grains to increase porosity in bricks. Construction and Building Materials. 2005;**19**:117-126. DOI: 10.1016/j.conbuildmat.2004.05.014

[28] Gromes R, Zeuch M, Piendl A. Further investigations into the dietary fibre content of beers. Brau International. 2000;**18**:24-28

[29] Lin TI. Content and variety of functional components in hops, beer and brewer's by-products [thesis]. Taipei: Fu Jen University; 2016

[30] Lewis MJ, Young TW. Brewing. London, UK: Chapman and Hall; 1995

[31] Bamforth CW. Nutritional aspects of beer-a review. Nutrition Research. 2002;**22**:227-237. DOI: 10.1016/S0271-5317(01)00360-8

[32] Gorjanovic S, Novakovic M, Potkonjak N, Leskosek-Cukalovic I, Suznjevic D. Application of a novel antioxidative assay in beer analysis and brewing process monitoring. Journal of Agricultural and Food Chemistry. 2010;**58**:744-751. DOI: 10.1021/jf903091n

[33] Bartolome B, Faulds CB, Williamso G. Enzymic release of ferulic acid from barley spent grain. Journal of Cereal Science. 1997;**25**:285-288. DOI: 10.1006/jcrs.1996.0091

[34] Stevens JF, Page JE. Xanthohumol and related prenylflavonoids from hops and beer: to your good health! Phytochemistry. 2004;**64**:1317-1330. DOI: 10.1016/j.phytochem.2004.04.025

[35] Stevens JF, Taylor AW, Deinzer ML. Quanitative analysis of xanthohumol and related prenylflavonoids in hops and beer : by liquid chromatography-tandem mass spectrometry. Journal of Chromatography A. 1999;**832**:97-107. DOI: 10.1016/S0021-9673(98) 01001-2

[36] Kao TH, Wu GY. Simultaneous determination of prenylflavonoid and hop bitter acid in beer lee by HPLC-DAD-MS. Food Chemistry. 2013;**141**:1218-1226. DOI: 10.1016/j.foodchem.2013.04.032

[37] Van Cleemput M, Cattoor K, Bosscher KD, Haegeman G, Keukeleire DD, Heyerick A. Hop (*Humulus lupulus*)-derived bitter acids as multipotent bioactive compounds. Journal of Natural Products. 2009;**72**:1220-1230. DOI: 10.1021/np800740m

[38] Shah BN, Panchal MA, Gohil N, Nayak BS, Modi DC. Phyto-pharmacol profile of *Humulus lupulus*. Pharmacology. 2010;**1**:719-736

[39] Haseleu G, Intelmann D, Hofmann T. Identification and RP-HPLC-ESI-MS/MS quantitation of bitter-tasting β-acid transformation products in beer. Journal of Agricultural and Food Chemistry. 2009;**57**:7480-7489. DOI: 10.1021/jf901759y

[40] Verzele M. 100 years of hop chemistry and its relevance to brewing. Journal of the Institute of Brewing.1986;**92**:32-48. DOI: 10.1002/j.2050-0416.1986.tb04372.x

[41] Blanco CA, Rojas A, Caballero PA, Ronda F, Gomez M, Caballero I. A better control of beer properties by predicting acidity of hop iso-α-acids. Trends in Food Science and Technology. 2006;**17**:373-377. DOI: 10.1016/j.tifs.2005.11.012

[42] Moir M. Hops: a millennium review. Journal of the American Society of Brewing Chemists. 2000;**58**:131-146. DOI: 10.1094/ASBCJ-58-0131

[43] Wu GY. Analysis of functional components in beer lee and its antiobesity activities [thesis]. Taipei: Fu Jen University; 2012

[44] Chuang ZL. Anti-obesity effect of brewer's yeast biomass in animal model [thesis]. Taipei: Fu Jen University; 2013

[45] Kiyofuji A, Yui K, Takahashi K, Osada K. Effects of xanthohumol-rich hop extract on the differentiation of preadipocytes. Journal of Oleo Science. 2014;**63**:593-597. DOI: 10.5650/jos.ess14009

[46] Yui K, Kiyofuji A, OK. Effects of xanthohumol-rich extract from the hop on fatty acid metabolism in rats fed a high-fat diet. Journal of Oleo Science. 2014;**63**:159-168. DOI: 10.5650/jos.ess13136

[47] Lin CY. Antiproliferative effect of ethanol extract from beer brewer's yeast biomass in lung cancer cell lines [thesis]. Taipei: Fu Jen University; 2013

[48] Lee IS, Lim J, Gal J, Kang JC, Kim HJ, Kang BY, Choi HJ. Anti-inflammatory activity of xanthohumol involves heme oxygenase-1 induction via NRF2-ARE signaling in microglial BV2 cells. Neurochemistry International. 2011;**58**:153-160. DOI: 10.1016/j.neuint.2010.11.008

[49] Tang W, Chen LH, Daun H, Ho CT, Pan MH. Inhibitory effects of hexahydro-β-acids in LPS-stimulated murine macrophage. Journal of Functional Foods. 2011;**3**:215-222. DOI: 10.1016/j.jff.2011.04.004

[50] Dorn C, Massinger S, Wuzik A, Heilmann J, Hellerbrand C. Xanthohumol suppresses inflammatory response to warm ischemia-reperfusion induced liver injury. Experimental and Molecular Pathology. 2013;**94**:10-16. DOI: 10.1016/j.yexmp.2012.05.003

[51] Saugspier M, Dorn C, Thasler WE, Heilmann J, Hellerbrand C. Hop bitter acids exhibit anti-fibrogenic effects on hepatic stellate cells *in vitro*. Experimental and Molecular Pathology. 2011;**92**:222-228. DOI: 10.1016/j.yexmp.2011.11.005

[52] Hougee S, Faber J, Sanders A, van den Berg WB, Garssen J, Smit HF, Hoijer MA: Selective inhibition of COX-2 by a standardized CO2 extract of *Humulus lupulus* L. in *vitro* and its activity in a mouse model of zymosan-induced arthritis. Planta Medica. 2006;**72**:228-233

[53] Pinto C, Duque AL, Rodríquez-Galdón B, Cestero JJ, Macías P. Xanthohumol prevents carbon tetrachloride-induced acute liver injury in rats. Food and Chemical Toxicology. 2012;**50**:3405-3412. DOI: 10.1016/j.fct.2012.07.035

[54] Dorn C, Heilmann J, Hellerbrand C. Protective effect of xanthohumol on toxin-induced liver inflammation and fibrosis. International Journal of Clinical and Experimental Pathology. 2012;**5**:29-36

[55] Gerhauser C, Alt A, Heiss E, Gamal-Eldeen A, Klimo K, Knauft J, Neumann I, Scherf H-R, Frank N, Bartsch H, Becker H. Cancer chemopreventive activity of xanthohumol, a natural product derived from hop. Molecular Cancer Therapeutics. 2002;**1**:959-969

[56] Liu Y, Gu X, Tang J, Liu K. Antioxidant activities of hops (*Humulus lupulus*) and their products. Journal of the American Society of Brewing Chemists. 2007;**65**:116-121. DOI: 10.1094/ASBCJ-2007-0211-01

[57] Namikoshi T, Tomita N, Fujimoto S, Haruna Y, Ohzeki M, Komai N, Sasaki T, Yoshida A, Kashihara N. Isohumulones derived from hops ameliorate renal injury via an anti-oxidative effect in Dahl salt-sensitive rats. Hypertension Research. 2007;**30**:175-184. DOI: 10.1291/hypres.30.175

[58] Milligan S, Kalita J, Pocock V, Heyerick A, De Cooman L, Rong H, De KD. Oestrogenic activity of the hop phyto-oestrogen, 8-prenylnaringenin. Reproduction. 2002;**123**:235-242. DOI: 10.1530/rep.0.1230235

[59] Christoffel J, Rimoldi G, Wuttke W. Effects of 8-prenylnaringenin on the hypothal-amo-pituitary-uterine axis in rats after 3-month treatment. Journal of Endocrinology. 2006;**188**:397-405. DOI: 10.1677/joe.1.06384

[60] Shamoto T, Matsuo Y, Shibata T, Tsuboi K, Takahashi H, Funahashi H, Okada Y, Takeyama H. Xanthohumol inhibits angiogenesis through VEGF and IL-8 in pancreatic cancer. Pancreatology. 2013;**13**:S52-S53. DOI: 10.1016/j.pan.2013.07.203

[61] Shimamura M, Hazato T, Ashino H, Yamamoto Y, Iwasaki E, Tobe H, Yamamoto K, Yamamoto S. Inhibition of angiogenesis by humulone, a bitter acid from beer hop. Biochemical and Biophysical Research Communications. 2001;**289**:220-224. DOI: 10.1006/bbrc.2001.5934

[62] Zanoli P, Zavatti M. Pharmacognostic and pharmacological profile of *Humulus lupulus* L. Journal of Ethnopharmacology. 2008;**116**:383-396. DOI: 10.1016/j.jep.2008.01.011

[63] Drenzek JG, Seiler NL, Jaskula-Sztul R, Rausch MM, Rose SL. Xanthohumol decreases Notch1 expression and cell growth by cell cycle arrest and induction of apoptosis in epithelial ovarian cancer cell lines. Gynecologic Oncology. 2011;**122**:396-401. DOI: 10.1016/j.ygyno.2011.04.027

[64] Festa M, Capsso A, D'Acunto CW, Masullo M, Rossi AG, Pizza C, Piacente S. Xanthohumol induces apoptosis in human malignant glioblastoma cells by increasing reactive oxygen species and activating MAPK pathways. Journal of Natural Products. 2011;**74**:2505-2513. DOI: 10.1021/np200390x

[65] Lamy V, Roussi S, Chaabi M, Gossé F, Schall N, Lobstein A, Raul F. Chemopreventive effects of lupulone, a hop β-acid, on human colon cancer-derived metastatic SW620 cells and in a rat model of colon carcinogenesis. Carcinogenesis. 2007;**28**:1575-1581. DOI: 10.1093/carcin/bgm080

[66] Yasukawa K, Takeuchi M, Takido M. Humulone, a bitter in the hop, inhibits tumor promotion by 12-0-tetradecanoylphorbol-13-acetate in two-stage carcinogenesis in mouse skin. Oncology. 1995;**52**:156-158. DOI: 10.1159/000227448

[67] Zanoli P, Zavatti M, Rivasi M, Brusiani F, Losi G, Puia G, Avallone R, Baraldi M. Evidence that the β-acids fraction of hops reduces central GABAergic neurotransmission. Journal of Ethnopharmacology. 2007;**109**:87-92. DOI: 10.1016/j.jep.2006.07.008

[68] Sanchez CL, Franco L, Bravo R, Rubio C, Rodríguez AB. Beer and health: benefits for sleep. Revista Española de Nutrición Comunitaria. 2010;**16**:160-163

[69] Yajima H, Ikeshima E, Shiraki M, Kanaya T, Fujiwara D, Odai H, Tsuboyama-Kasaoka N, Ezaki O, Oikawa S, Kondo K. Isohumulones, bitter acids derived from hops, activate both peroxisome proliferator-activated receptor α and γ and reduce insulin resistance. Journal of Biological Chemistry. 2004;**279**:33456-33462. DOI: 10.1074/jbc.M403456200

[70] Miura Y, Hosono M, Oyamada C, Odai H, Oikawa S, Kondo K. Dietary isohumulones, the bitter components of beer, raise plasma HDL-cholesterol levels and reduce liver cholesterol and triacylglycerol contents similar to PPARα activations in C57BL/6 mice. British Journal of Nutrition. 2005;**93**:559-567. DOI: 10.1079/BJN20041384

[71] Tobe H, Muraki Y, Kitamura K, Komiyama O, Sato Y, Sugioka T, Maruyama HB. Bone resorption inhibitors from hop extract. Bioscience, Biotechnology, and Biochemistry. 1997;**61**:158-159. DOI: 10.1271/bbb.61.158

[72] Ding C, Parameswaran V, Udayan R, Burgess J, Jones GJ. Circulating levels of inflammatory markers predict change in bone mineral density and resorption in older adults: A longitudinal study. The Journal of Clinical Endocrinology and Metabolism. 2008;**93**:1952-1958. DOI: 10.1210/jc.2007-2325

Chapter 5

Assessment of Bioaccessibility: A Vital Aspect for Determining the Efficacy of Superfoods

Abstract

Bioaccessibility is a vital aspect when qualifying food products to be superfoods. It could be defined as the amount of a food constituent which is present in the gut, as a consequence of its release from the solid food matrix. This chapter highlights the evaluation of the bioaccessibility of three studies using an *in vitro* model of digestion, involving potential superfoods which are as follows: (1) Five Sri Lankan endemic fruits (2) ten spices which are commonly used in culinary preparations in Sri Lanka as well as throughout the world, and (3) three Kombucha 'tea fungus' fermented beverages obtained through different microbial cultures. In all three studies, the antioxidant and starch hydrolase activities of the food products were evaluated, given the therapeutic importance of these characteristics. In the first study, it was observed that the antioxidant and activities had decreased, although the starch hydrolase inhibitory activities had been sustained. The remaining two studies demonstrated that both these functional properties had statistically significantly increased ($P < 0.05$), or sustained. Overall, the studies emphasise the need for evaluating the bioaccessibility of functional properties of potential superfoods before they can be properly advocated for consumption to the general public.

Keywords: antioxidants, bioaccessibility, bioavailability, functional food, starch hydrolase inhibitory activity

1. Introduction

Superfoods have been thought to possess many health benefits mostly owing to the existence of bioactive compounds. Most of the superfoods which are believed to possess disease preventive properties have demonstrated superior antioxidant activity. This property itself is one of the most coveted mechanisms touted to prevent the occurrence of non-communicable diseases such as cardiovascular disease (CVD), diabetes and cancer. Carotenoids and phenolic compounds are the two major categories of dietary antioxidants where over a hundred member compounds have been identified per each class to date. Carotenoids are fat-soluble, mostly occurring in the form of colouring agents and pigments [1]. On the other hand, phenolic compounds exist as free, esterified, etherified and insoluble-bound forms and are commonly found in edible fruits and vegetables, leafy vegetables, roots, tubers, bulbs, herbs, spices and legumes [1]. A recent trait of these compounds which is of relevance to superfoods is their ability to inhibit starch hydrolases – in particular, α-amylase and α-glucosidase. Starch hydrolase inhibitory activity prevents the sudden release of glucose into the physiological system, thereby preventing the triggering of biochemical pathways which produce free radicals inside the mitochondria as a result of glucose metabolism. Since the digested products of α-amylase act as substrates for α-glucosidase, inhibition of α-amylase is believed to be more important in preventing the commencement of the biochemical pathway of starch digestion and thereby, curbing the release of glucose to the physiological system.

When it comes to bioactive compounds in superfoods – antioxidants in particular, their release and availability from the food matrix into the digestive tract is of paramount importance. This is technically referred to as bioavailability. One step prior to bioavailability is bioaccessibility. This is the amount of a food constituent which is present in the gut, as a consequence of its release from the solid food matrix, and thus, which may be able to pass through the intestinal barrier into the circulatory system. Considering both bioaccessibility and bioavailability, only bioactive constituents released from the food matrix by the digestive enzymes are bioaccessible in the gut, and therefore are potentially bioavailable. Digestive enzymes play an important role when it comes to bioaccessibility, since their action may essentially increase or decrease their release into the digestive system. As a consequence of this phenomenon, it is evident that the amount of bioaccessible food antioxidants and other therapeutic compounds of interest may differ quantitatively and qualitatively from the values which have been included in food databases.

Several *in vitro* and *in vivo* methodologies have been developed over the past few years to assess the bioaccessibility of bioactive compounds in many of the superfoods [1, 2]. *In vivo* methods using animal models or in the form of clinical trials, usually provide the most accurate results. However, the two major shortcomings of *in vivo* assessments is that they are time consuming and costly. Thus, much effort has been devoted to the development of *in vitro* procedures, where nearly accurate outcomes can be achieved within a short span of experimental time. However, it has to be borne in mind that any *in vitro* method would eventually fall short in matching the precision which can be achieved by actually studying the behaviour of a food *in vivo* due to the natural complexity of the release and digestive process itself [2]. As a result of this, some compromise is understandably needed between the accuracy and ease of utilisation of any *in vitro* digestion model which aims at assessing bioaccessibility. During the past few years, food and animal scientists have utilised several *in vitro* digestion models to test the structural and chemical changes occurring in different food under simulated gastrointestinal (GI) conditions. However, none of these methods have yet been widely accepted and no method has been singularly standardised in terms of assessment.

This chapter aims at giving examples demonstrating the importance of assessing bioaccessibility of the antioxidant and starch hydrolase inhibitory through three types of food products subjected to pancreatic and duodenal digestion: (1) Five Sri Lankan endemic fruits (2) ten spices which are commonly used in culinary preparations in Sri Lanka as well as throughout the world, and (3) three Kombucha 'tea fungus' fermented beverages obtained through different microbial cultures. Although not entirely confirmed, all these food products have demonstrated to possess significant health benefits, mostly owing to the phenolic compounds present in them and their corresponding antioxidant activities. The starch hydrolase inhibitory activities of some of these food products have not been systematically explored before given the native nature of the plant products. The results enclosed herewith show the importance of measuring the total antioxidant capacity and starch hydrolase inhibitory properties before and after pancreatic and duodenal digestion using an *in vitro* model, where further studies based on the outcomes can be conducted on some of the products.

2. Stability of the antioxidant and starch hydrolase inhibitory activities of five Sri Lankan fruits subjected to pancreatic and duodenal digestion

As a tropical country, Sri Lanka has been gifted with a myriad of fruits some of which are rare and endemic and consumed by locals for generations for their associated therapeutic properties. Given their extensive and historical usage as traditional medicines, it is to be expected that many of these fruits are superfoods, although this aspect has not been systematically evaluated and confirmed to date.

In order to verify these functional properties, the stability of the antioxidant and starch hydrolase inhibitory activities of the following fruits were evaluated when subjected to pancreatic and duodenal digestion: *Elaeocarpus serratus* (ES), *Flacourtia indica* (FInd), *Flacourtia inermis* (FIne), *Pouteria campechiana* (PC), *Solanum nigrum* (SN). The digestion model consisted of exposure of the macerated pulps of these fruits to the enzymatic action of pepsin and pancreatin following the model of Wootton-Beard et al. [2]. The total phenolics content (TPC), Oxygen Radical Absorbance Capacity (ORAC), di(phenyl)-(2,4,6-trinitrophenyl) iminoazanium (DPPH) radical scavenging and α-amylase & α-glucosidase inhibitory activities of the fruit pulps, prior to digestion as well as following pancreatic and duodenal digestion were evaluated. The TPC was determined according to Singleton et al. [3] and expressed in milligrammes gallic acid equivalents per gram (mg GAE/g). The ORAC and DPPH radical scavenging activities were determined according to Prior et al. [3] and Brand-Williams et al. [5] respectively, and expressed in μmol trolox equivalents per gram (μmol TE/g) and EC_{50} (mg/kg), respectively. The starch hydrolase inhibitory activities – both α-amylase and α-glucosidase, were determined according to Liu et al. [6]. The results were expressed for this particular parameter in terms of acarbose equivalents per gram (AE/g), where acarbose is a known starch hydrolase inhibitor prescribed to diabetic patients.

Changes to the antioxidant activities are shown in **Figure 1**, while the starch hydrolase inhibitory activities are shown in **Figure 2**. ES was observed to possess the highest TPC, ORAC and DPPH radical scavenging activities prior to digestion. Following both digestion phases, all antioxidant activities had statistically significantly decreased ($P < 0.05$) in the fruit pulps. A higher correlation was observed between TPC and ORAC ($R^2 = 0.872$), as compared with TPC

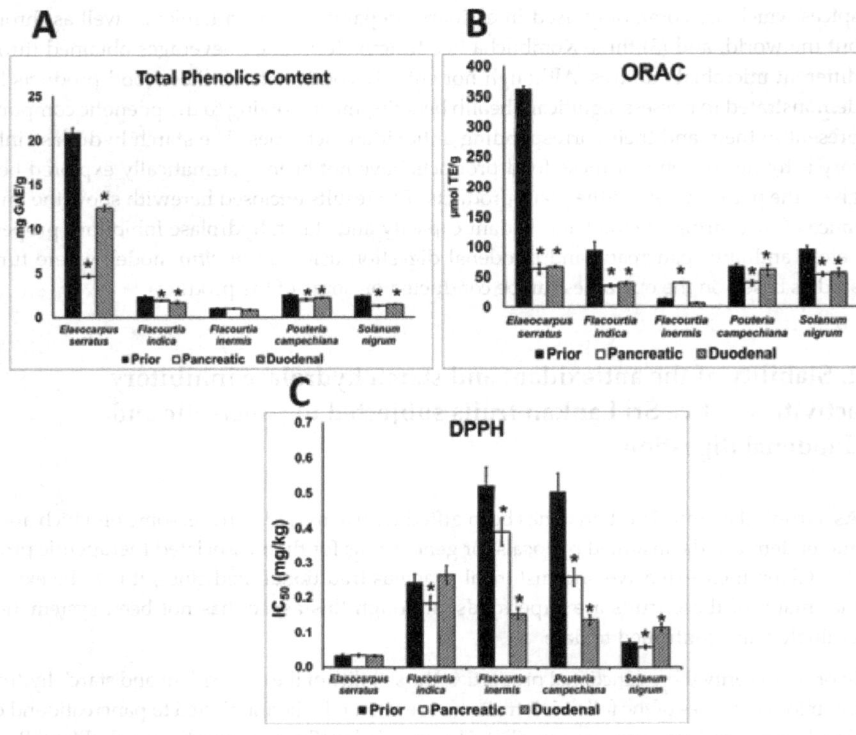

Figure 1. The (A) TPC, (B) ORAC and (C) DPPH radical scavenging activities of the fruits, prior to digestion as well as following exposure to pepsin and pancreatin enzymatic action. $P < 0.05$ denotes statistically significant difference as compared with the respective values prior to *in vitro* digestion. Values represent mean ± SEM of 3 ≤ independent experiments.

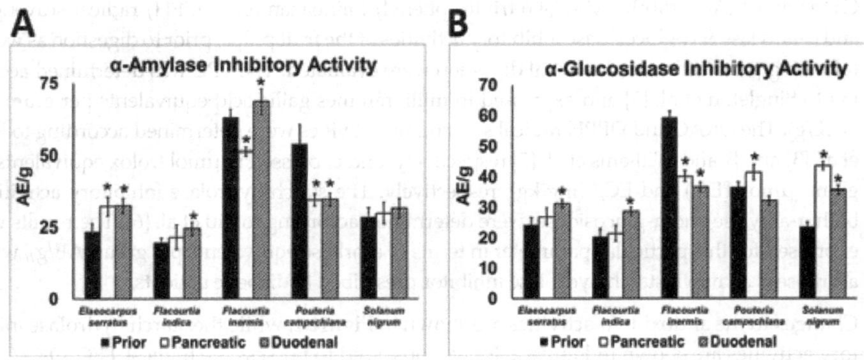

Figure 2. The (A) α-amylase and (B) α-glucosidase inhibitory activities of the fruits, prior to digestion as well as following exposure to pepsin and pancreatin enzymatic action. $P < 0.05$ denotes statistically significant difference as compared with the respective values prior to *in vitro* digestion. Values represent mean ± SEM of 3 ≤ independent experiments.

and DPPH radical scavenging activities ($R^2 = 0.341$), indicating that the phenolic compounds of the fruits were better scavengers of hydroxyl radicals which are generated during the assay. FIne was observed to possess the highest starch hydrolase inhibitory activities prior to digestion. Statistically significant ($P < 0.05$) enhancements in this characteristic were observed in all samples following duodenal digestion. There was no observable correlation between the TPC and starch hydrolase inhibitory activities, which implied that this particular characteristic may not have been necessarily extended from the phenolic compounds.

The mechanistic action behind the reduction in TPC, ORAC and DPPH radical scavenging activities would have to be further explored based on these outcomes. In fact, many of the previous researches examining fruit juices in a similar nature have consistently shown large decreases in total antioxidant capacities (TAC) post digestion [7–9]. One study which demonstrates otherwise is that of Wootton-Beard et al. [2]. With reference to the TPC, it was suggested by Ryan et al. [10] that this may have been due to a structural transformation of polyphenols which render them undetectable to be assessed via the assay methodology. However, the reduction in the antioxidant activities strongly suggests that the functional properties of the fruits maybe hampered when subjected to the digestion process.

Overall, the study highlights the necessity of obtaining biologically relevant information on antioxidants by providing data concerning the bioaccessibility and bioavailability as well for that matter, in a human system. The model by Wootton-Beard et al. [2] could essentially be a key methodology in this aspect, in order to obtain preliminary data, prior to embarking on *in vivo* and clinical studies. It must also be mentioned that the antioxidant and starch hydrolase inhibitory activities of the five endemic Sri Lankan fruits are reported herein for the first time. There are many more fruits which are native to the country which have not been studied in similar manner, and given the outcome of this study, it appears to be imperative that they be investigated for their potentials as superfoods.

3. Evaluation of the antioxidant and starch hydrolase inhibitory activity of 10 commonly used household spices subjected to *in vitro* digestion

Spices and herbs are generally used for flavouring and colouring purposes when it comes to culinary preparations. However, they are also known for their medical or antiseptic properties – characteristics which have been owed to the presence of immense amounts of antioxidant compounds. They have been particularly touted as superfoods based on their antioxidant properties. For this particular study, the following spices were selected for assessing the stability of their antioxidant and starch hydrolase inhibitory properties based on their frequent usage in Sri Lanka households as well as throughout the world – any previous studies carried out on their therapeutic potentials of these spices were taken into account during the selection as well: Cardamom (CA – *Elettaria cardamomum*), cloves (CV – *Syzygium aromaticum*), coriander (CO – *Coriandrum sativum*), cumin seeds (CS – *Cuminum cyminum*), curry leaves (CL – *Murraya koenigii*), fenugreek (FG – *Trigonella foenum*), mustard (MU – *Brassica nigra*), nutmeg (NM – *Myristica fragrans*), sweet cumin (SC – *Pimpinella anisum*), and star anise (SA – *Illicium verum*). In order to reduce the content of moisture, powders of these spices were obtained in dried form from the Ayurvedic Medicinal Hall in Kandy, Sri Lanka. The methodology by Wu et al. [11] was followed for the preparation of the herbal extracts. For the digestion procedure, the same methodology as Wootton-Beard et al. [2] was used. In addition to the TPC, DPPH and starch hydrolase inhibitory activities mentioned in the previous study, quantification of the water-soluble as well as oil-soluble ORAC values was carried out. The water-soluble ORAC ($ORAC_{FL}$) was carried out according to the method by Prior et al. [4], while the lipophilic ORAC ($ORAC_{oil}$) was carried out according to the method by Hay et al. [12]. Both parameters were expressed as μmol TE/g. The rationale for using both these ORAC methodologies for this study was because spices are known to possess many oil-soluble antioxidant pigments – in particular, carotenoids, along with phenolic compounds. Thus, it could be deemed necessary to assess the antioxidant activity extending from these oil-soluble as well as water-soluble antioxidant compounds. Also, the FRAP assay was carried out as described by Benzie and Strain [13], while the $ABTS^+$ radical scavenging activity was evaluated using a methodology previously reported by Ozgen et al. [14]. These two assays were added to this study as compared with the previous one, in order to capture the antioxidant activities extending from the various methods of scavenging radicals. Results from these two particular assays were reported as μmol TE/g. Similar to the previous study, the starch hydrolase inhibitory activities were determined according to Liu et al. [6], where the results were expressed as AE/g.

Changes to the TPC, $ORAC_{FL}$, $ORAC_{oil}$, FRAP, DPPH and $ABTS^+$ radical scavenging activities are shown in **Figure 3**, while the starch hydrolase inhibitory activities are shown in **Figure 4**. When taking the TPC values prior to digestion, CV had the highest amount (22.83 ± 0.20 mg GAE/g) followed by CL (21.94 ± 0.19 mg GAE/g). NM had the lowest TPC (0.80 ± 0.03 mg GAE/g). Following the gastric phase of digestion, all spice extracts had statistically significant increases ($P < 0.05$) in terms of the TPC. This trend was observed in the duodenal phase of digestion as well ($P < 0.05$). As compared with the previous study on the endemic fruits, it could be said that the digestion process released more phenolic compounds, thus increasing their bioaccessibility.

Despite the increases in the TPC in both digestion phases, the $ORAC_{FL}$, ABTS, DPPH and FRAP results did not indicate similar trends, although a few exceptions were observed. Nevertheless, it was heartening to observe that the antioxidant activity values coming from these assays had either been maintained during the digestion phases or statistically significantly increased ($P < 0.05$). This was another difference as compared with the study on the endemic fruits. Similar to the $ORAC_{FL}$ values, the $ORAC_{oil}$ values did not display any statistically significant increases or decreases ($P < 0.05$) as compared with the values prior to the gastric and duodenal digestion phases. The only exception in this instance was CA, where a statistically significant increase ($P < 0.05$) was observed in the duodenal digestion phase. Although the $ABTS^+$, DPPH and FRAP assay values followed an almost similar trend as the $ORAC_{FL}$ values, their correlation with the TPC was comparatively less. A clear correlation between each of the antioxidant assays used to evaluate the TAC was also not observed.

When it comes to the starch hydrolase inhibitory activities, FG had the highest α-amylase and α-glucosidase inhibitory activities, while CS had the lowest for both enzymes. With the exception of CS, none of the spice extracts showed statistically significant changes ($P < 0.05$) to the initial enzyme inhibitory values prior to being exposed to gastric and duodenal digestion. This observation was of significance since the initial starch hydrolase inhibitory activities of the spice extracts were maintained even though they were exposed to the digestive enzymes. The spices were observed to inhibit α-amylase better than α-glucosidase, based on the mean inhibitory values. This characteristic is of significance as well. Inhibition of α-amylase is considered

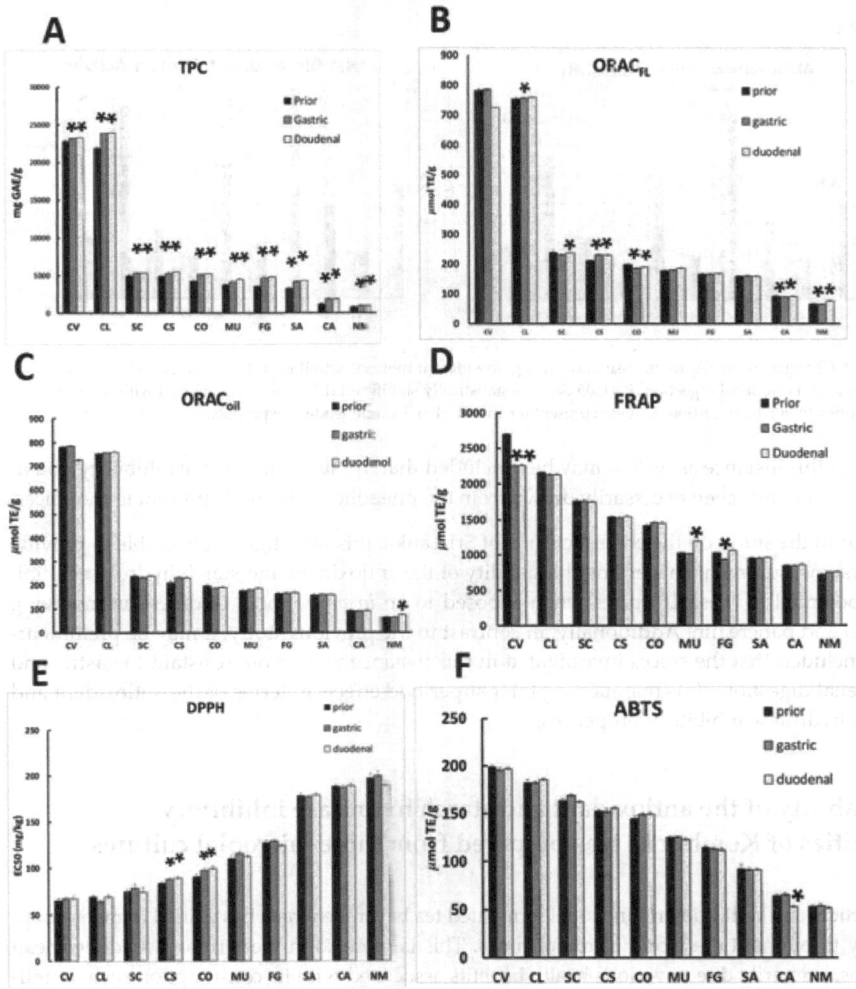

Figure 3. Changes to the (A) TPC, (B) ORAC$_{FL}$, (C) ORAC$_{oil}$, (D) FRAP, (E) DPPH and (F) ABTS^{+} radical scavenging activities of the 10 spices when subjected to pancreatic and duodenal digestion. $P < 0.05$ denotes statistically significant difference as compared with the respective values prior to *in vitro* digestion. Values represent mean ± SEM of 3 ≤ independent experiments.

to be more important when it comes to reducing the breakdown of starch, since as explained before, it triggers the production of the substrate for the subsequent action of α-glucosidase. Thus, the spices demonstrated the capability to inhibit the cascade of enzymatic starch breakdown as a whole. Similar to the study on the endemic fruits of Sri Lanka, a clear correlation between the starch hydrolase inhibitory activities and TPC was not observed in this study.

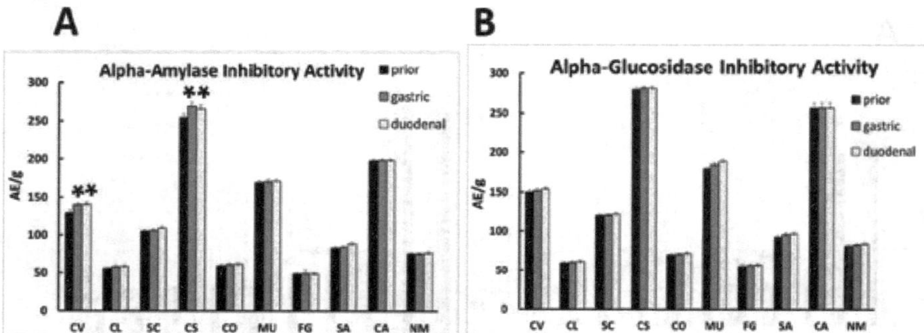

Figure 4. Changes to the (A) α-amylase and (B) α-glucosidase inhibitory activities of the 10 spices when subjected to pancreatic and duodenal digestion. $P < 0.05$ denotes statistically significant difference as compared with the respective values prior to *in vitro* digestion. Values represent mean ± SEM of 3 ≤ independent experiments.

Thus, in this instance as well, it may be concluded that the starch hydrolase inhibitory potential may not have been necessarily drawn from the phenolic compounds present in the spices.

Similar to the study on the endemic fruits of Sri Lanka, this investigation was able to provide the first measurement concerning the stability of the antioxidant and starch hydrolase inhibitory potential of these 10 spices when exposed to an *in vitro* model of digestion involving pepsin and pancreatin. Additionally, in contrast to the previous study, it may be preliminarily concluded that the spices investigated in this instance were more resistant to gastric and duodenal digestion, thus maintaining their superfood effects in terms of the antioxidant and starch hydrolase inhibitory properties.

4. Stability of the antioxidant and starch hydrolase inhibitory activities of Kombucha teas prepared from three microbial cultures

Kombucha is a well-known functional fermented tea beverage which has gained immense popularity throughout the world in recent times. This is more so in Western and Mediterranean regions, primarily due to various health benefits associated with its consumption such as anticancer, anti-diabetic, anti-inflammatory, hepatoprotective and detoxification properties as well as its ability to act as a probiotic [15]. The studies focusing on the antioxidant activity of the beverage have also demonstrated the possession of various bioactive ingredients of therapeutic interest, mainly polyphenols and other categories of secondary metabolites which are generated as a result of the fermentation process itself [16]. The bacterial component of the Kombucha culture consists of strains such as *Acetobacter xylinum, A. xylinoides, A. aceti, A. pausterianus,* and *Bacterium gluconicum,* while the dominant yeast strains are *Zygosaccharomyces bailii, Schizosaccharomyces pombe, Saccharomyces ludwigii, S. cerevisiae, Kloeckera* spp., *Torulaspora* spp., and *Pichia* species [17]. Given its purported health benefits, it could be easily seen that Kombucha is a potential superfood and thus, its bioaccessibility should be evaluated.

This study evaluated the changes of the antioxidant activity, TPC and the starch hydrolase inhibitory activity in sugared black tea fermented with three Kombucha cultures possessing differed microbial compositions, when subjected to pancreatic and duodenal digestion. The bioaccessible antioxidant capacity and starch hydrolase inhibitory activities were evaluated, similar to the two previous studies on the endemic fruits of Sri Lanka and the spice extracts. Whether sufficient antioxidant and starch hydrolase inhibitory properties were released from the fermented beverage and subsequently released for absorption during the intestinal and duodenal digestion processes was of interest and worth investigating. Study of such phenomenon would be of value, especially to demonstrate the applicability of *in vitro* digestion models for fermented functional beverages, and also to observe whether the functional properties can be sustained during the pancreatic and duodenal digestion phases.

Three Kombucha tea fungal mats were used in this study and their microbial compositions in terms of the dominant species were determined as follows:

- K1: *Acetobacter aceti, Zygosaccharomyces bailii, Brettanomyces claussenii*
- K2: *A. aceti,* Saccharomyces ludwigii, *Zygosaccharomyces rouxii*
- K3: *A. aceti, Lactobacillus* spp., *Leuconostoc* spp., *S. ludwigii*

Three separate portions of sugared black tea were inoculated with 3% (w/v) of each of the cultures above aseptically for 7 days at 24 ± 3°C. This fermentation time has been identified as the ideal duration prior to unwanted metabolites being formed as a result of extended periods of microbial activity. The fermented broth was subjected to the *in vitro* digestion process detailed by Wootton-Beard et al. [2]. The TPC, ORAC and DPPH radical scavenging activities were measured using the same methods mentioned in the two previous studies on the endemic fruits and spice extracts. However, for this study, the superoxide radical scavenging activity was used. This was because previous studies had used this assay for this particular beverage with noteworthy outcomes, indicating that it has a superior ability to scavenge superoxide radicals [15, 16]. The value was expressed in terms of percentage inhibition (%). The starch hydrolase inhibitory activities were also evaluated in this study using the same methodology described by Hay et al. [12]. However, given that the samples in this study contained ongoing microbial activities, the α-amylase and α-glucosidase inhibitory activities were expressed as IC_{50} (µg/mL) instead of AE/g.

Changes to the TPC, ORAC, DPPH and superoxide radical scavenging activities are shown in **Figure 5**, while the starch hydrolase inhibitory activities are shown in **Figure 6**. Statistical comparisons were done with the fermented beverage at day 7, prior to being subjected to the *in vitro* digestion process. When the three fermented teas were subjected to the *in vitro* digestion phases, statistically significant increases ($P < 0.05$) were observed in all after intestinal digestion. In addition, statistically significant increases ($P < 0.05$) were observed in the TPC following the duodenal digestion phase as compared with the undigested sample at day 7 of the fermentation as well. As for the antioxidant values, ORAC value for all teas prior to digestion was within a range of 2460–2640 μmol TE/mL. K3 had the highest ORAC value among all three fermented samples by the end of the fermentation process. In comparing the values during the digestion processes, statistically significant increases ($P < 0.05$) were observed following both phases in all three beverages. Similar to the study on the spices, an increase in the ORAC value represents the sustenance of the antioxidant potential when exposed to pancreatic and duodenal digestion and this could be of therapeutic importance. The increase in the ORAC value of K3 was less than K1 and K2 in the duodenal digestion phases as compared with the undigested counterparts. Nevertheless, the final ORAC values of all three teas were within a comparable range at the end of the two digestion phases.

The DPPH EC_{50} values of the three fermented Kombucha teas remained within the range of 56–59 mg/kg prior to the digestion phases. In contrast to the ORAC assay values, there were no statistically significant changes ($P < 0.05$) to the DPPH EC_{50} values and the superoxide scavenging activities when subjected to the digestion process. This was noteworthy, considering the instance of the study on endemic fruits. It was apparent that the fermented beverages had retained its antioxidant potential in all three beverages, despite the enzymatic activities of pepsin and pancreatin. A better correlation between the TPC and the ORAC values were observed in comparison to the correlation between DPPH EC_{50} and superoxide scavenging values. This

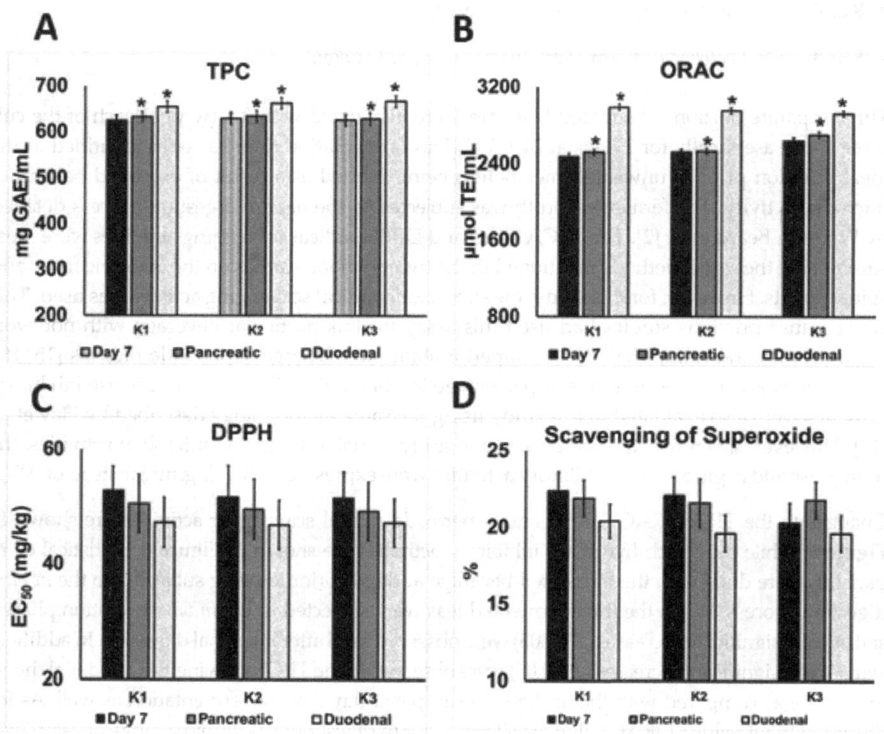

Figure 5. Changes to the (A) TPC, (B) ORAC, (C) DPPH scavenging activity and (D) superoxide scavenging activity of the three fermented beverages. $P < 0.05$ denotes statistically significant difference as compared with the respective values prior to *in vitro* digestion. Values represent mean ± SEM of 3 ≤ independent experiments.

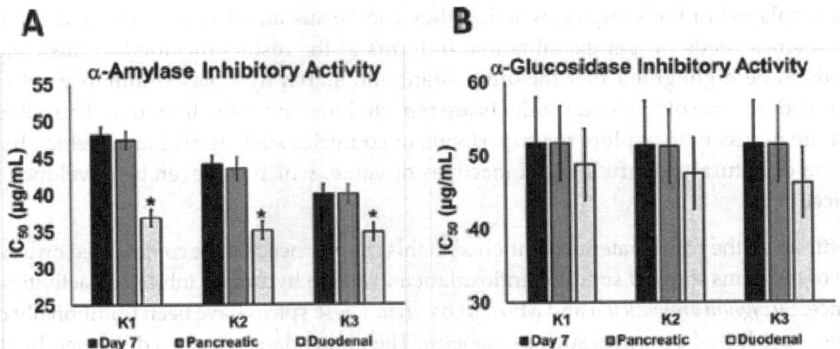

Figure 6. Changes to the (A) α-amylase and (B) α-glucosidase inhibitory activities of the three fermented Kombucha samples when subjected to pancreatic and duodenal digestion. $P < 0.05$ denotes statistically significant difference as compared with the respective values prior to *in vitro* digestion. Values represent mean ± SEM of 3 ≤ independent experiments.

maybe because the phenolic compounds present in all three types of tea samples possess a better scavenging activity of peroxide radicals which are generated during the ORAC assay.

There were notable observations in terms of the starch hydrolase inhibitory activities. When considering the results obtained for the three Kombucha strains used in this study, it was observed that the α-glucosidase inhibitory activity of K3 after duodenal digestion was higher than the other two beverages. Additionally, when the three beverages were subjected to pancreatic digestion, the α-amylase inhibitory activity remains more so the same compared with the undigested counterpart for all three teas. However, after duodenal digestion the α-amylase inhibitory activity displayed statistically significant increases ($P < 0.05$) compared with the control sample (i.e. undigested beverage at 7 days of fermentation). Statistically significant changes ($P < 0.05$) in the α-glucosidase inhibitory activity were not observed in any of the teas. This was of importance, since the starch hydrolase inhibitory activities demonstrated stability against the enzymatic activity of pepsin and pancreatin.

In conclusion, the tea fermented with the K3 pellicle was discovered to be the better Kombucha beverage in terms of having the highest antioxidant and starch hydrolase inhibitory activities following fermentation, as well as its resistance to enzymatic activity of pepsin and pancreatin. Since the K3 Kombucha sample contains *Lactobacillus* spp., this beverage can be used as a potentially good probiotic supplement and could also be considered as a prospective superfood being superior in terms of antioxidant potential and probiotic effects.

5. Conclusions

In conclusion, this chapter details the application of an *in vitro* digestion system to a variety of food samples – mostly of plant origin, and measurement parameters. It is obvious from the outcomes that despite superior antioxidant and starch hydrolase activities being displayed in the samples, whether they can be sustained when subjected to digestive enzymes needs proper examination. In terms of the results included in this chapter, it needs to be highlighted that the antioxidant and starch hydrolase inhibitory activities mentioned in some of the food products are reported herein for the first time. Thus, it highlights the necessity to explore for superfoods in countries such as Sri Lanka which have a plethora of natural resources with bioactives of value, which can even be developed into nutraceuticals.

The efficacy of the plant material mentioned in this chapter needs to be commented on as well. Some of the items showed superior antioxidant and starch hydrolase inhibitory activities, for instance, *Syzygium aromaticum* and *Murraya koenigii*. These spices have been traditionally used for many medicinal purposes such as diabetes. The antioxidant activities displayed by these spices only affirmed their potential in combating hyperglycemia-induced oxidative stress. Nevertheless, the actual efficacy of these plant material and the fermented beverages could only be confirmed if they were subjected to clinical evaluations.

Several studies have utilised *in vitro* digestion methods to analyse structural changes, bioavailability, bioaccessibility and digestibility of foods, indicating that *in vitro* digestion systems are common useful tools for analyses of the efficacy of foods and drugs. In terms of bioaccessibility, there is clearly an urgent need for more research into *in vitro–in vivo* correlations with well-defined systems, so that more realistic *in vitro* models can be developed for screening purposes. Finally, between bioavailability and bioaccessibility, bioavailability has been given more attention; however, from the studies highlighted in this chapter, it is apparent that bioaccessibility deserves equal attention, since it is a precursor to bioavailability, and also an important aspect which determines whether a food product is indeed a superfood.

References

[1] Garrett DA, Failla MA, Sarama RJ. Development of an *in vitro* digestion method to assess carotenoid bioavailability from meals. Journal of Agricultural and Food Chemistry. 1999;**47**(10):4301-4301. DOI: 10.1021/jf9903298

[2] Wootton-Beard PC, Moran A, Ryan L. Stability of the total antioxidant capacity and total polyphenol content of 23 commercially available vegetable juices before and after *in vitro* digestion measured by FRAP, DPPH, ABTS and Folin–Ciocalteu methods. Food Research International. 2011;**44**(2011):217-224. DOI: 10.1016/j.foodres.2010.10.033

[3] Singleton VL, Orthofer R, Lamuela-Ravent'os RM. Analysis of total phenols and other oxidation substrates and. Methods in Enzymology. 1998;**299**:152-178

[4] Prior RL, Hoang H, Gu L, et al. Assays for hydrophilic and lipophilic antioxidant capacity (oxygen radical absorbance capacity (ORACFL)) of plasma and other biological and food samples. Journal of Agricultural and Food Chemistry. 2003;**51**(11):3273-3279

[5] Brand-Williams W, Cuvelier ME, Berset C. Use of a free-radical method to evaluate antioxidant activity. Food Science and Technology-Lebensmittel-Wissenschaft & Technologie. 1995;**28**(1):25-30

[6] Liu T, Song L, Wang H, Huang DJ. A high-throughput assay for quantification of starch hydrolase inhibition based on turbidity measurement. Journal of Agricultural and Food Chemistry. 2011;**59**(18):9756-9762

[7] Bermudez-Soto MJ, Tomas-Barberan FA, Garcia-Conesa MT. Stability of polyphenols in chokeberry (*Aronia melanocarpa*) subjected to *in vitro* gastric and pancreatic digestion. Food Chemistry. 2007;**102**(3):865-874

[8] McDougall GJ, Dobson P, Smith P, Blake A, Stewart D. Assessing potential bioavallability of raspberry anthocyanins using an *in vitro* digestion system. Journal of Agricultural and Food Chemistry. 2005;**53**(15):5896-5904

[9] McDougall GJ, Fyffe S, Dobson P, Stewart D. Anthocyanins from red wine – Their stability under simulated gastrointestinal digestion. Phytochemistry. 2005;**66**(21):2540-2548

[10] Ryan L, Prescott SL. Stability of the antioxidant capacity of twenty-five commercially available fruit juices subjected to an *in vitro* digestion. International Journal of Food Science and Technology. 2010;**45**(6):1191-1197

[11] Wu X, Beecher GR, Holden JM, Haytowitz DB, Gebhardt SE, Prior RL. Lipophilic and hydrophilic antioxidant capacities of common foods in the United States. Journal of Agricultural and Food Chemistry. 2004;**52**(12):4026-4037

[12] Hay KX, Waisundara VY, Timmins M, et al. High-throughput quantitation of peroxyl radical scavenging capacity in bulk oils. Journal of Agricultural and Food Chemistry. 2006;**54**(15):5299-5305

[13] Benzie IFF, Strain JJ. The ferric reducing ability of plasma (FRAP) as a measure of 'antioxidant power': The FRAP assay. Analytical Biochemistry. 1996;**239**(1):70-76

[14] Ozgen M, Reese RN, Tulio AZ, Scheerens JC, Miller AR. Modified 2,2-azino-bis-3-ethylbenzothiazoline-6-sulfonic acid (ABTS) method to measure antioxidant capacity of selected small fruits and comparison to ferric reducing antioxidant power (FRAP) and 2,2-diphenyl-1-picrylhydrazyl (DPPH) methods. Journal of Agricultural and Food Chemistry. 2006;**54**(4):1151-1157

[15] Mindani IW, Jayawardena N, Gunawardhana CB, Waisundara VY. Health, wellness, and safety aspects of the consumption of Kombucha. Journal of Chemistry. 2015;**2015**:1-11. DOI: 10.1155/2015/591869

[16] Jayabalan R, Marimuthu S, Swaminathan K. Changes in content of organic acids and tea polyphenols during kombucha tea fermentation. Food Chemistry. 2007;**102**(1):392-398

[17] Marsh AJ, O'Sullivan O, Hill C, Ross RP, Cotter PD. Sequence-based analysis of the bacterial and fungal compositions of multiple kombucha (tea fungus) samples. Food Microbiology. 2014;**38**:171-178

Index

A

Adipocytes 3
Agronomical processes 17
Akt protein 24
Allithiamine 4
Alzheimer's disease (AD) 5,7,11,23,32
Angiogenesis 24
Anthocyanin 2,4,23,24
Anti-cancer effects 5
Anti-cancer ingredients 5
Antidiabetic activity 23
Antioxidant defense systems 13
Arteriosclerosis 3
Atherosclerosis 15

B

Bifidobacteria 4
Bioactive compounds 34
Biological membranes 20
Blood brain barrier 6
Blood circulation 4
Blood–brain barrier 24
Brain functions 5
Brown adipocytes 3

C

Caloric restriction 4
Capsaicin 3
Cardiovascular diseases (CVDs) 15,23,34
Cardiovascular factors 14
Catechins 3
Central Nervous System (CNS) 18
Chronic hypertension 20
Chronic inflammation 23
Circadian cycle 18
Circadian rhythm 21
Cognitive functions 14
Cultivar conditions 17
Cyclooxygenase (COX-2) 23

D

Degenerative diseases 23
Degrading triglycerides 3
Dietary fiber 33,41,48,53,77
Dietary sources 17
Dietary supplement 1
Dipropyl trisulfide (DPTS) 6
Docosahexaenoic acid (DHA) 3,15,34
Domino effect 2

E

Eat foods 1
Eicosapentaenoicacid (EPA) 3,15
Endogenous hydroxytyrosol 21

F

Fat oxidation 37
Fatty acids 3
Fish processing 34
Fish protein 35
Fish proteinhydrolysates (FPH) 35
Fish protein isolates (FPI) 35
Fish protein powder (FPP) 35
Fucoidan 5
Functional foods 1

G

Genestin4,5,7
Glial cells 24
Global status 33

Glucose 3
Glutamatergic neurotoxicity 13

H

Health benefits 17,23
Healthful foods 33
Heart disease 1
High blood pressure 2
Hippocampus 24
Homeostasis 4,9,36
Honeycomb 17
Human organism 34
Hydrogen peroxide 15
Hydronium (H3O+) 36
Hydrophobic 36
Hydroxytyrosol 13,21,23
Hypercholesterolemia 14
Hyperlipidemia 2

I

Ideal diet 2,6,7
Immunesystem diseases 15
Indolamine 19
Insulin-degrading enzyme (IDE) 6
Ischemic diseases 14
Isoelectric solubilization/
precipitation(ISP) 33,35

L

Lipopolysaccharide 23

M

Market trend 33
Mediterranean diet6,13,14
Melatonin 13,18,19,23
Melatonin molecule 18
Metabolic syndrome 2
Mitochondria 3
Mitochondrial dysfunction 14
Mitochondrial fragmentation 24
Myofibrillar protein 35

N

N-1-Acetyl-N-2-formyl-5-
methoxykynuramine (AFMK) 20
National Cancer Institute 5
Neprilysin (NEP) 6
Neural network 5
Neurodegenerative diseases 13,14,24
Neurohormone 18
Neuroinflammatory processes 13
Neuroprotection 16
Neuroprotective compound 23
Nitric oxide synthase (NOSe) 24
Nutritious protein 33
Nutritive qualities 34

O

Oleic acid 3
Oleuropein 21
Olive oil 3
O-methylation 18
Oxidase 23
Oxidative stress 13
Oxygenspecies 13

P

Parkinson disease 13,18
Pathological level 14
Pharmacokinetics 16
Phospholipids 5
Pinealgland 18
Polyphenols 3
Polyunsaturated fatty acids (PUFA)3,15
Postpone neurodegenerative diseases 17
Prevent cancer 4
Proanthocyanidins 23,24
Proapoptotic proteins 14

R

Reactive oxygen species (ROS) 4
Resveratrol 16
Rotenone neurotoxicity 24

C

Salt substitute 33
Serotonin 18
Sesaminol 4
Shitake 5
Sirtuin gene 4
Sulforaphane 2
Superfood 1
Synaptic signaling 24

T

TCA cycle 3
Traditionalnutrients 34
Triglyceride 3
Trophic factors 14
Tryptophan 18

U

Ubicuitin-proteasome system 14
Uncoupling protein 1 (UCP1) 3

V

Veratrumgrandiflorum 16
Vitis vinifera 16

W

Wakame 3,5

X

Xanthone 2

Z

Zero electrostatic charge 36
Zwitterions 36